Techniques
with Bacteria

Rosa K. Pawsey

Hutchinson Educational

Hutchinson Educational Ltd
3 Fitzroy Square, London W1

London Melbourne Sydney Auckland
Wellington Johannesburg Cape Town
and agencies throughout the world

First published 1974
©Rosa K. Pawsey 1974
© Illustrations Hutchinson Educational Ltd 1974

Printed in Great Britain by
The Anchor Press, and bound by
Wm. Brendon, both of Tiptree, Essex

ISBN 0 09 115611 4

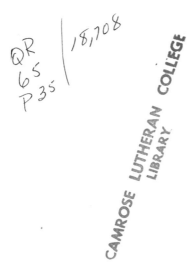

Contents

Acknowledgements

I would like to thank the several people who have helped me, and given me advice at various stages in the preparation of the manuscript. I was greatly helped in the production of photographs by Dr Malcolm Cobb of Ponthir, Monmouthshire; and also Mr Alan Reynolds and Mr Alan Rangnekar of the National College of Food Technology, University of Reading.

1 Simple equipment and its uses

The simple equipment of micro-biology can be divided into two categories:

   i. containing equipment or vessels
   ii. transfer equipment.

Many of the items in both categories can be found as part of the equipment in the average school laboratory.

   Sections at the end of the chapter describe miscellaneous equipment and consider the arrangement of the laboratory for micro-biology.

Containing Equipment or Vessels

Vessels have several uses.
   for containing media before and after sterilisation
   for the growth of cultures
   for the observation of chemical or immunological reactions
All vessels must fulfil certain experimental requirements. For example, if a vessel is to be used to incubate a culture its size and shape must provide adequate room for growth, no extraneous organisms must be able to enter, and, preferably, it must enable the experimenter to observe the culture. In addition the vessel must be made from materials which will withstand the rigours of sterilisation, or else be disposable. The inside surface must be easy to clean and must not be rough or of awkward shape. Experience will soon enable you to know which pieces of equipment fulfil your experimental criteria.

i. Tubes

Test tubes There are several types which may be used but they should be of bacteriological quality, that is capable of withstanding sterilisation. The most common type in use has a round base, and a rimless top. There is less risk of the top of the test tubes being broken when many are packed together in bundles if tubes are rimless. *Uses:* For containing media before and after sterilisation, for the growth of cultures, for the observation of chemical reactions.

Flat bottomed test tubes have the advantage that they can stand up on their own but offer no other special advantage over the normal round based test tube. *Uses:* These are the same as for the previous type of tube.

Durham's tubes are small, round bottomed test tubes of similar size to combustion tubes (internal diameter about 0·7 cm, length about 2·5 cm) *Use:* For detecting the production of gas by organisms in fermentation tests.

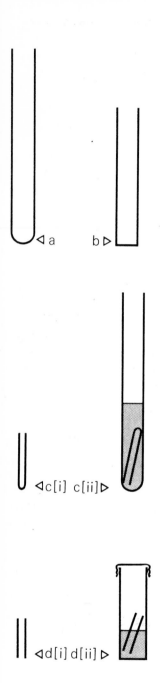

Fig. 1
Showing the different types of tubes.
a] round based tube
b] flat based tube
c] i Durham's tube
   ii Durham's tube in use
d] i Craigie tube
   ii Craigie tube in use.

Precipitin tubes These are small tubes of internal diameter about 0·5 cm and length about 5·0 cm and having a special tapered base. *Use:* For observing immunological reactions. It is unlikely that a school would require these.

Craigie tubes are open ended, cut from a piece of glass tubing of internal diameter 0·7 cm and about 5·0 cm long. *Use:* They are used in a cultural method for detecting the motility of organisms. (See p. 80)

Test tube racks are preferably made from metal so that they can be sterilised in an autoclave or with disinfectant. Try to avoid wooden racks.

Test tube baskets are wire mesh baskets designed to hold about 5 dozen tubes and used in the autoclave.

ii. Flasks

Round or conical flat bottomed flasks with short necks are used for making media and as storage vessels for media. They are also used as culture vessels when fairly large quantities of an organism are required. Ordinary laboratory flasks can be used provided that they are well cleaned and washed out so that they contain no trace of chemicals inside.

2·5 litre, 1 litre and 250 cm³ flasks are used.

Long necked flasks similar to those used in volumetric analysis are used rarely, if at all.

Stoppers for flasks will be discussed in the section 'Stoppers' later in this chapter.

iii. Petri dishes

These are flat circular dishes made of glass or metal which have a base and a lid. A culture of micro-organisms is grown on a thin layer of nutrient poured into the base of the dish. The base is approximately 10 cm in diameter and 1·5 cm in depth; the lid is similar in shape but slightly larger so that it makes a loose fit over the base. Their design allows for the free exchange of gases between the inside and the outside but minimises the chances of contamination when a culture is set up in the dish. Petri dishes can be stacked in piles for sterilisation, storage or incubation. There are several types:

Pressed glass: The glass is thick and has a long life but in time its surface gets scratched so that the transparency is obscured. These very heat resistant dishes do not stack very well. They are fairly cheap to buy.

Boro-silicate glass: Blown petri dishes whose glass does not scratch and which stack well. They are fairly delicate and also expensive, the risk of breakages, which could be fairly high in a school, might obviate their use.

Metal lids are made of aluminium and can be used in conjunction with glass bases. They would cut down the breakages but have the disadvantage that, since they are opaque, cultures must be viewed with the lids removed. Aluminium bases can also be bought.

Polystyrene petri dishes are much lighter than glass and can be bought in sealed packets of 10 or 12 already sterilised. They are excellent and where only a small number of experiments are to be done or small numbers of pupils are involved, could be the solution to many of the practical problems facing the teacher. They can prove to be expensive but are less so when bought in large numbers.

Points to remember when buying petri dishes: Tough petri dishes should withstand the life they may suffer at the hands of children. Petri dishes which stack well will save space. Metal lids are cheap, reduce breakages, stack well, and are useful for displaying other biological specimens.

Sterilisation cans for petri dishes are circular cans of copper or stainless steel which hold one dozen petri dishes. They are packed with clean dry dishes, the lid put on and the whole is placed in the hot air oven for sterilisation. Afterwards the can is not opened until the dishes are to be used. These cans are very useful but are expensive. A 3kg biscuit tin will serve the same purpose as well.

Improvised petri dishes: Several containers from the home can be used as substitute petri dishes, provided that the conditions of the experiment are not too exacting.

iv. Bottles

Should be made of glass and in such a way that they will withstand the rigours of sterilisation. There are two sizes in common use:

The universal bottle: This is straight sided with a wide neck and has a rubber washer in the aluminium screw cap. It has a total capacity of 24 cm^3.

The McCartney bottle is similar to, but has slightly less capacity than, the universal bottle, and is narrow necked.

The bijou bottle: This has a similar shape to the universal bottle but has a total capacity of 8cm^3.

Bottles are used in the following ways:

as containers for growing organisms in liquid or solid culture,

as containers for stock cultures,

for the storage of small quantities of sterile medium, in which media do not dry out and the risk of contamination is very low. Bottles are sterilised in the autoclave because the rubber washers in the screw caps cannot withstand sterilisation temperatures in the hot air oven.

v. Beakers

Are not often used in micro-biology. No special type is required however they are sometimes used as water baths.

vi. Lids and stoppers

All vessels have lids to obviate contamination.

Cotton wool plugs: Use cotton wool 'bacteriological quality' which is non-absorbent. This will filter the air and stop the passage of micro-organisms into the vessel. White and coloured plugs can be used for stoppers to identify different media. Cotton wool plugs are made in a special way so that when a plug is removed from the top of a test tube or flask it retains its shape and is easy to reinsert:

Method 1: A strip of cotton wool slightly wider than the mouth of the vessel and three times the desired finished length is pulled off the roll. The strip is folded into three, and the smooth end, *S*, is inserted into the mouth of the vessel. (See Fig. 2.)

Method 2: A roughly square piece of cotton wool is taken and folded lengthwise into three. The centre of the roll is then pushed, using a glass rod, into the top of the vessel so that two-thirds of the stopper is in and one-third is out of the vessel. (See Fig. 3.)

It is a matter of practice to make a stopper of the right size. It should retain its shape when removed from a vessel, it should be easy to re-insert, and after re-insertion should not expand and push itself out.

Cotton wool stoppers have the disadvantage that they singe easily on hot glass. After use they are either incinerated or thrown away into disinfectant.

Metal caps: Loose fitting aluminium caps can be bought for test tubes and for flasks. They are easy to use and can be used time and time again. They withstand sterilisation and are supplied in many colours which is useful because each colour can be used to denote the contents of the vessel.

They do tend after a time to change in size, so that caps may become either too big or small and no longer fit. The colours also eventually fade.

Screw caps for bottles: It is important to remember to loosen screw caps when sterilising bottles in the autoclave, and that the rubber liners do not withstand sterilisation in the hot air oven.

Aluminium foil caps: Domestic aluminium foil can be used to make disposable caps for test tubes and flasks. This foil does tend to tear so great care should be taken. It would be advisable to fasten such caps to flasks by tying them with a piece of string (do not use rubber bands—these will perish in the hot air oven.)

Transfer Equipment

The following instruments are used to transfer micro-organisms from one medium to another:

i. loops and wires

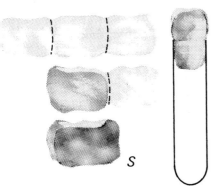

Fig. 2
Showing method No. 1 for making cotton wool stoppers

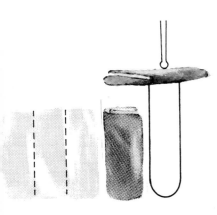

Fig. 3
Showing method No. 2 for making cotton wool stoppers

    ii. glass spreaders

    iii. pipettes

(Refer to Chapter 4 'Cultures' to the ways in which these instruments are used.)

i. Loops and wires

Small wire loop: A length of wire fixed into a metal or glass handle at one end and bent into a loop at the other end. Originally platinum wire was used but nowadays a nichrome (nickel-chromium) or 'Eureka' wire (S.W.G. 22; diameter 0·5mm) is used. The wire plus loop is approximately 8cm long, not including the length required to fix it into the handle. The loop should have a diameter of 2·5—3·0mm and should be bent so that the loop is in line with the main axis of the wire (*not* at right angles). The handle should be about 20cm long. The loop must be closed so that a film of liquid can form across the loop. *Use:* This type of loop is used to transfer a small inoculum from liquid or solid media to other media and for spreading agar plates. Wire loops are part of the essential equipment. (See Fig. 4.)

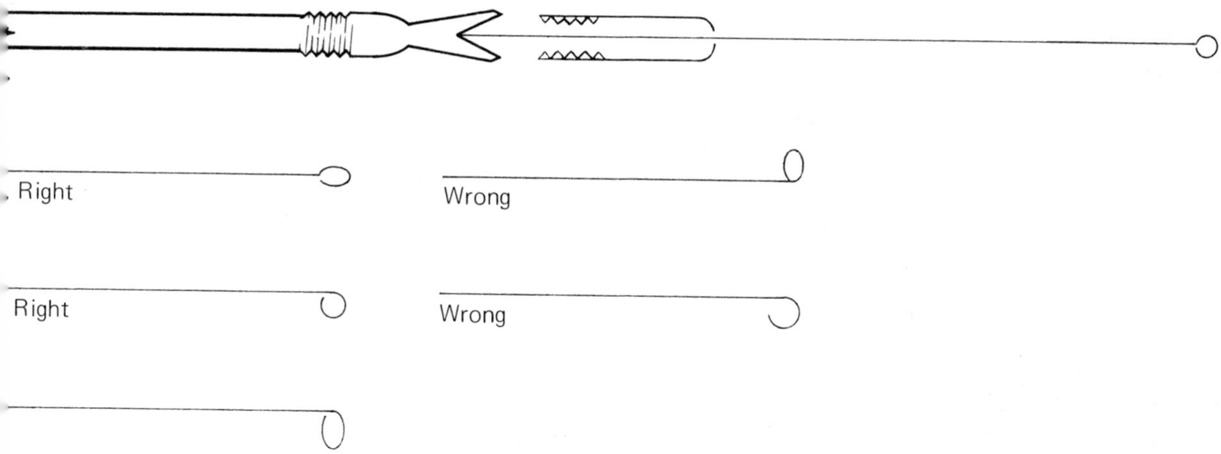

Right   Wrong

Right   Wrong

Fig. 4

owing points about the structure small and large wire loops.

section of the screw chuck of a loop.

the orientation to the long axis of the wire for the small loop—right and wrong.

shape of the small loop—right and wrong.

orientation of the large loop to the main axis of the wire.

Large wire loop: This is very similar to the small wire loop except in two respects. The loop is larger being 5-6mm in diameter and is bent at right angles to the main axis of the wire (c.f. small loop). *Use:* It is used where an inoculum is to be superimposed on to a plate culture. 'Spot inoculation' (Refer to Chapter 4, 'Cultures')

Straight wire: A loop straightened out will serve very well as a straight wire. It is used if a very small part of a bacterial colony is to be removed, or for making stab cultures (See Chapter 4, 'Cultures')

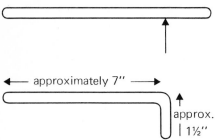

Fig. 5
Showing the dimensions of a glass spreader

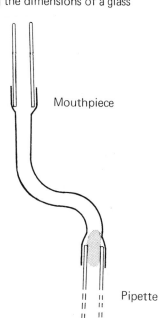

Mouthpiece

Pipette

Fig. 6
Showing a mouth piece and its attached tubing

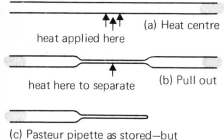

(a) Heat centre
heat applied here

(b) Pull out
heat here to separate

(c) Pasteur pipette as stored—but used with a teat

Fig. 7
Showing the stages in making a Pasteur pipette

ii. Glass spreaders

A length of glass rod 3-4mm diameter and about 9 inches long is heated at a point about 1½ inches from its end and a right angle bend made at that point. Both ends of the rod are heated and smoothed off so that there are no rough or sharp edges. A glass spreader is used to obtain confluent growth on a plate inoculated with a large inoculum (See 'Cultures').

iii. Pipettes

Mouth pieces: A mouth piece is simply a short piece of glass tubing (2 inches—3 inches) whose ends have been smoothed and to which a length of rubber tubing is attached. The other end of the rubber tubing is attached to the pipette. The glass tube is clamped between the premolars on one side of the mouth in such a position that the end can be sealed at will with the side of the tongue. A MOUTH PIECE IS ALWAYS USED IN CONJUNCTION WITH A PIPETTE TO AVOID ANY RISK OF SUCKING MICRO-ORGANISMS UP INTO THE MOUTH. THIS IS AN IMPORTANT SAFETY PRECAUTION AND SHOULD ALWAYS BE OBSERVED.

Pasteur pipettes: A length of sterile glass tubing of internal diameter 0·6 cm and 25 cm in length (approximately) is heated midway along its length and drawn out to give two Pasteur pipettes. (See 'Storage'). The capillary ends are heat sealed. It is important that the capillary part should be kept straight in line with the axis of the rest of the pipette. The wide end of a Pasteur pipette is stopped with a small plug of cotton wool, made in the same way as for use in test tubes. The plug acts as a bacterial filter. It is left in position when the pipette is being used and therefore cuts down the risk of sucking organisms into the mouth. When a sterile Pasteur pipette is to be used the sealed end of the capillary tube is broken off with sterile forceps. The finished dimensions should be approximately: wide bore 11 cm, narrow bore 9 cm (total length approximately 20 cm). *Use:* These pipettes are used for transferring small unmeasured quantities of liquid media or cultures.

Storage of Pasteur pipettes must be such that once they are sterilised they will not get accidentally broken or contaminated. It is a good idea to cut 25 cm lengths from stock glass tubing, to plug the ends and to heat sterilise these tubes in a metal container. The sterile undrawn tubes can be drawn and used as required.

Alternatively the tubes can all be drawn at once, sterilised, and the stock of pipettes kept in their tin resting on a bed of cotton wool. If they are kept vertically the wide bore is uppermost with the capillaries resting on a thick bed of cotton wool. The reason for this is that the outside as well as the inside of the capillary comes into contact with sterile media, cultures etc. and must therefore itself be kept sterile. The pipettes are handled at the wide end only.

A third alternative is to draw the pipettes, seal the capillary end, stop the other with cotton wool, wrap each individually in Kraft paper and then sterilise. Each remains wrapped until it is used. [Kraft paper can be purchased from any scientific supplier].

After use the contaminated pipette is transferred directly into a bowl or dish containing disinfectant. The whole pipette should be immersed and allowed to soak for 24 hours. Or they can be placed in a water bath and boiled. (Refer to Chapter 3, 'Sterilisation', for details). After sterilisation they can be used again but they must have a new cotton wool stopper and the capillary must be re-sealed.

Other pipettes: Graduated 1 cm³, 2 cm³, 5 cm³, and 10 cm³, pipettes are used on occasions when measured quantities are required. They are used in exactly the same way as Pasteur pipettes and observing the same precautions. They can be stored in the same ways and are, after use treated in the same way.

iv. Forceps

Metal forceps are used on many occasions. They can be sterilised by dipping them in alcohol and flaming them or by boiling. Take care that the alcohol does not run down and burn your hand too. (Refer to Chapter 3).

Sections 1 and 2 of this chapter have described the basic equipment which is necessary to contain micro-organisms and their media, and also the equipment needed to transfer them from place to place. However there are a few other items rather miscellaneous in character which form part of the basic simple equipment.

Miscellaneous Equipment
i. Grease pencils

Clear labelling of all cultures, tubes and so on which are involved in all experiments is vital. It is not advisable to use sticky labels for labelling because there is distinct risk of the transference of organism into the mouth. (See page 37, 'Aseptic technique'). Instead, grease pencils are used for most labelling jobs. They write well on clean glass and later the marks can easily be removed with xylene. Black and other colours are cheaply available from laboratory suppliers. Grease pencils write best if the surface is grease free and is slightly warm. They tend to break easily so it is better not to sharpen them to a fine point. A piece of rubber tubing over the end of the pencil will prevent the pencil rubbing off on your laboratory overall. Grease pencils are indispensable.

ii. Slides

A fair quantity of glass slides will be needed if you intend to stain micro-organisms and view them under the microscope. Glass slides can be used again and again but they must be thoroughly cleaned between uses.

Cavity slides: It may be useful for demonstration purposes to have a few—say half a dozen—cavity slides, but they are not essential equipment. (See Chapter 5, 'Staining and making micro-organisms visible'.)

iii. Cover slips

It is essential to have a supply of cover slips, but not necessary to have a large number. 3/4 or 7/8 of an inch square, number 1 thickness is best. Thicker ones may prevent focussing under high power (with oil immersion).

iv. Solutions and reagents

Distilled water: Some solutions are made with distilled water. It can be dispensed from polythene wash bottles.

Saline solution: 0·9% w/v solution is used on occasion as a diluent.

Immersion oil: Oil of the same refractive index as glass and which does not harden is used with an oil immersion lens to view stained smears. The oil is dispensed either from glass bottles with a glass rod, or from polythene 'squeezee' bottles. A small quantity, 50 cm^3, will last a long time.

Lysol or other proprietary disinfectant: It is absolutely essential that the stock of disinfectant is maintained and is not allowed to run out. During *all* practical work with micro-organisms each work bench in the laboratory should have two or three jam jars or other similar receptacles full of diluted disinfectant. In the event of a culture being dropped or splashed immediate disinfection can be carried out. Care should be taken in the use of lysol since it can have a burning effect on the skin. If the skin is splashed it should be washed promptly with running water.

Arrangement of the Laboratory

Micro-biological work can be done in any reasonable school laboratory. It is not necessary to have a laboratory which resembles an aseptic room in a hospital.

The benches should be of polished wood or laminated plastic or other substance which presents a smooth surface and which can be disinfected by swabbing. Gas points for bunsen burners should be available on the work benches. There should at least be one sink with a tap per bench—preferably more. In addition there should be a hot water supply in the laboratory and next to this, soap and disposable towels. Lastly, in common with normal practice there should be a good first aid set.

Summary

The simple equipment necessary to carry out micro-biological experiments is divided into categories:

> *containers:* tubes, flasks, petri dishes, bottles and their lids, cotton wool.
> *transfer equipment:* loops and wires, spreaders, pipettes and mouth pieces, forceps.
> *miscellaneous:* grease pencils, slides, cover slips, solutions and reagents, disinfectant.

Guidance is given to general laboratory layout.

2 Making, and using media —

Experimental work in micro-biology consumes media and when classes or year groups are large the teacher is confronted with the problems of quantities, costs and time.

This chapter explains the theory of media and the different methods by which media can be made or obtained. Each teacher must eventually assess his or her own problems in order to conclude which approach suits best.

1. Introduction—liquid and solid media

For the growth of any particular organism:

    i. the medium must supply all essential nutritional requirements.

    ii. moisture must be available to the organism. (Gelling agents sequester water so that care must be taken that they are not added in quantities greater than necessary, otherwise water will not be available to the organism for growth)

    iii. the pH must be adjusted so that it is in the range tolerated by the organism. (All micro-organisms have an optimum pH for growth.) For practical purposes, unless exacting experiments are being carried out the pH of the media for bacteria is usually adjusted to approximately 7·5 and for moulds to 5·4. pH can be tested using pH papers.

If these criteria are satisfied then the medium will support good growth provided that the conditions of incubation are correct.

There are two types of media: liquid and solid. Liquid media are used extensively for 'boosting' growth. This is because the convection currents in a liquid disperse the organisms and prevent overcrowding and the accumulation of the toxic end products of metabolism. They are used also for the observation of certain biochemical reactions. Two examples of liquid media are peptone water and nutrient broth.

However, a liquid medium alone may not satisfy the needs of the experiment. It is, for example, difficult or impossible to separate individual species which are growing in mixed culture in liquid medium, if only liquid media are available. Workers in the early days of micro-biology had to contend with this difficulty, which has now been overcome by the use of solid media. On solid culture medium the organisms remain where they were originally deposited, reproduction of the cells takes place, so that at the site of deposition of one cell its progeny form a pile of cells which soon becomes visible and is known as a colony. Each bacterial species forms a colony with a characteristic appearance and by which it can be recognised.

This means that if an inoculum from a mixed liquid culture is spread thinly over the surface of a solid medium (See Chapter 4, 'Cultures') in due course isolated colonies will become apparent. It is then a simple matter to remove a small piece of an isolated colony, inoculate that into a suitable sterile medium and obtain a pure culture of that particular species.

So solid media facilitate the separation of mixed cultures. They have many other uses, too numerous to be listed here but which will be referred to throughout the book.

Solid media are composed of two parts: the gelling bases and the nutrients.

i. Gelling bases

There are three types: gelatin, agar agar and silica gels.

Gelatin is a material which in the early days was used to solidify nutrient media. Today it is not used very much for this purpose because it has several disadvantageous properties. In 15% solution it forms a gel with water which liquefies at about 24°C. This means that it cannot be incubated and remains solid at temperatures around and above that. In addition it is a protein and may therefore be liquefied by the proteolytic activity of some organisms. In fact today gelatin is mainly used in the identification of organisms—'the gelatin liquefaction test'. Care must be taken when gelatin containing media are sterilised because prolonged exposure to high temperature denatures it, after which it will not gel on cooling.

Agar-agar: As a result of the difficulties experienced with gelatin, micro-biological workers looked for another gelling agent which suffered none of its disadvantages, and agar was found. It is a derivative of certain types of seaweed and is a long chain polysaccharide (d-galactopyranose) together with some impurities. Agar-agar can be purchased in either powder form or in fibre form. Mixed with water it must be heated to 100°C and kept at that temperature for a short while to bring it to solution. In 2% solution it sets at approximately 48°C. The gel once formed must be reheated to about 100°C before it will melt.

Agar is the most commonly used gelling agent because of its useful properties. In contrast to gelatin it can be incubated at all temperatures which support life, and remain a gel. (*Note*. The melting and solidifying temperatures given above may vary slightly from batch to batch because it is a natural product. The manufacturers do try to standardise as much as is possible.) Agar is inert to practically all species of organism.

Silica gels: As these are also inert to all organisms and because they supply no nutrients, silica gels are used to solidify media in experiments designed to ascertain the minimum nutritional requirements for organisms, and for the growth of autotrophic organisms, that is, those from the soil. Apart from the fact that they are expensive there is no reason why they should not be used in place of agar. Gelling agents must not be added in too high quantity to liquid media because high concentrations withdraw water from the medium preventing it from being available for the micro-organisms.

ii. Nutrients

When nutrients are incorporated into a gelling base the result is a solid nutrient medium.

2. Methods

The first thing to consider is the size of vessel in which the medium is to be made (See below). The second thing to consider is whether the glass is chemically clean, and if not to clean it (See page 21) and thirdly, of the ways available for making or obtaining media which is the best in your circumstances (See page 21).

Container Size

Media can be made in small or large quantities and can often conveniently be sterilised in the same container as the one in which they are made.

Table 1 showing quantities of media which can safely be held in various containers during sterilisation.

Container	Size	Quantity it can safely hold
Test Tube, rimless	5 inches x 0·5 inches	up to 4 cm^3 medium
	5 inches x 5/8 inches	up to 5-10 cm^3 medium
	6 inches x 3/4 inches	up to 10-15 cm^3 medium
	7 inches x 1 inch	up to 20 cm^3 medium greater than 20 cm^3 use a flask or bottle
Flask	250 ml	up to 100 cm^3 medium
	1 litre	up to 400 cm^3 medium
	2½ litre	up to 1 litre medium
Bottle	24 cm^3 (universal)	10 cm^3
	8 cm^3 (bijou)	2-3 cm^3

Table 1 gives a list of container sizes and the maximum volume of medium which they can contain for safe sterilisation. These volumes allow for expansion without explosion or overflow.

Table 2 indicates useful volumes in which media are often used.

Table 2 Useful volumes in which media are often used.

Liquid media	
Peptone water for biochemical tests	$5\,cm^3$ in a test tube
Peptone water sugars	$5\,cm^3$ in a test tube
Nutrient broth or other broths for cultures	$10\,cm^3$ in a test tube or screw cap bottle
Cooked meat media for culture of anaerobes	$20\,cm^3$ in screw cap bottle
Solid media	
Slopes	$5\,cm^3$ in test tube or bottle (though more can be used) $2\text{-}3\,cm^3$ in a bijou bottle
Plates	$12\text{-}15\,cm^3$ in a test tube for immediate use. In a screw cap bottle if to be stored before use.

Cleaning glassware

i. Glassware already containing cultures and to be used again, clean in the following way.

Cultures should be discarded into 3% lysol and left for 24 hours (to kill all organisms) *or* immersed in boiling 5% soap solution in soft or distilled water and boiled 15-20 minutes *or* autoclaved 15lb/15 minutes. Then clean well with a test tube brush in soft water and rinse in hot and cold water finally washing in distilled water. Drain and dry (in a hot air oven if available.) It is better not to rely on chemical disinfection but to use it preceding heat sterilisation.

ii. Glassware which has not previously been used for microbiology should be chemically cleaned first. Make a stock dichromate-sulphuric acid cleaning solution:

Dissolve 63 gm sodium or potassium dichromate in $35\,cm^3$ water by heating. Cool and add sulphuric acid up to 1 litre. Handle this corrosive liquid with care. If you splash your skin or clothes with it wash immediately in lots of water, and neutralise any residual acid with sodium carbonate solution.

Fill glassware with the cleaning solution and leave them standing for several days. Return the cleaning solution to its stock container. Wash glass 6 times with tap water and 3 times with distilled water.

Ways of obtaining media

There are several ways in which media can be obtained.

Starting with the raw ingredients is a very good way of making media but can be time consuming. Most of the media that one is likely to use can be obtained in the form of (dehydrated) granules, the manufacturers supply detailed instructions on the way to reconstitute the media. Some media can be bought in the form of tablets, which means that making media is very simple, and again the manufacturers give detailed instructions on reconstitution. For example to make 10 cm³ of nutrient broth one tablet is put into a universal bottle, 10 cm³ of distilled water added, the tablet allowed to dissolve, the lid put on the bottle and the whole sterilised. Some media can be bought made, sterilised and ready to use. Although it is unlikely that a school would have enough money to buy all its media in this way, it is worth remembering because unusual media that are difficult to make and which are used comparatively rarely can be bought like this.

Media from raw materials

Media prepared from fresh materials are of high quality but take time to prepare. The following methods are suggestions for media which might be made fairly readily by the teacher possibly with some help from the pupils. *Warning:* Do not use copper containers to make media or to contain media, because copper even in small quantities inhibits the growth of micro-organisms.

Liquid Media

Nutrient broth This is a very useful liquid medium which is richer than peptone water. It is a mixture of commercial peptone, meat extract, sodium chloride and tap water. The meat extract is made by holding finely divided lean meat in boiling water for a short while. Soluble constituents pass into solution and give the unconcentrated extract which is later concentrated. This contains gelatin, albumoses, peptones, amino acids and other nitrogen compounds. It also contains accessory growth factors (more than does peptone) and certain carbohydrates. In some media meat extract may be substituted by yeast extract (in whole or in part), and ordinary peptone by casein hydrolysate. *Use:* Nutrient broth will support the growth of most micro-organisms and a supply of it is essential for basic micro-biology.

Composition

Commercial bacteriological peptone	10gm
Meat extract	10gm
Sodium chloride	5gm
Tap water	1 litre

Mix these ingredients and allow them to dissolve in the water.
Adjust the pH to 7·5-7·6 with NaOH or HCl. If necessary filter through several layers of filter paper to remove precipitates. Dispense into the containers in which it is to be sterilised and

stored. Loosen caps of screw cap bottles and sterilise by
steaming 100°C/90 minutes.
or   autoclave 121°C/15 minutes.
(Refer to Chapter 3, 'Sterilisation')

Yeast extract (or Marmite, or Yeastrel) contains a range of amino acids, growth factors especially of the vitamin B group, inorganic salts and some carbohydrates. It may be used as a substitute for meat extract.

Glucose broth Nutrient broth is enriched by the addition of 1% glucose. Glucose acts as a reducing agent so this medium is used for the growth of anaerobic organisms. *Sterilise:* Tyndallisation in a steamer, or autoclave 10lb/15 minutes.

Peptone water. Commercial peptone is the water soluble product obtained from lean meat or other protein material, by digestion with proteolytic enzymes. The important constituents are peptones, proteoses, amino acids, a variety of inorganic salts and some accessory growth factors. The manufacturers supply different grades for different purposes: bacteriological, mycological, etc. Peptone water is peptone, plus sodium chloride dissolved in water. *Use:* Since peptone water contains no fermentable carbohydrate it is used as the basis medium for 'peptone water sugars', the media used for specific sugar fermentation tests. It is also used as the medium for other biochemical tests, and as a culture medium, though for good growth of many species it may need to be supplemented (See Table 3).

Composition
Peptone   1%
Sodium chloride   0·5%

Dissolve the ingredients in warm water. Filter. Adjust the pH to 7·2 to 7·4. *Sterilise:* Autoclave 15lb/15 minutes.
The golden granular form of peptone is highly hygroscopic so, when using it keep the lid off the bottle for the minimum length of time, and complete the weighing quickly.

Table 3 Sugars used in sugar fermentation tests

Monosaccharides
  (a) pentoses
  arabinose   — from gum acacia; and from sugar beet boiled with
                   dilute sulphuric acid
  rhamnose   — obtained by hydrolysis of quercitrin (from dyers oak)

(b) hexoses

glucose	— dextrose or grape sugar
fructose	— or laevulose—from many plants; formed in the inversion of cane sugar
mannose	— made from the ivory nut
galactose	— made by the hydrolysis of lactose

Disaccharides

sucrose	— saccharose, or cane sugar
maltose	— malt sugar
lactose	— milk sugar
trehalose	— from ergot, and several species of yeasts and fungi

Trisaccharides

raffinose	— from cotton seed meal, and sugar beet residues

Polysaccharides

starch	— soluble starch is usually prepared from potato starch
inulin	— from dahlia tubers
dextrin	— made by the partial hydrolysis of starch
glycogen	— from the livers of mammals and lower animals, also occurs in yeast and lower animals

Meat infusion broth
Fresh lean beef or ox heart
Commercial peptone
Sodium chloride
Carefully free the meat of its fat. Mince the lean meat as finely as possible. Add tap water in the proportion of 500 gm meat to one litre of water. Allow extraction to take place over 24 hours at a low temperature, in a refrigerator. Then strain it through muslin. Keep the bright red fluid which contains soluble meat proteins. Remove any fat from the surface by skimming it with a piece of filter paper. Then boil it for 15 minutes. It will become brown and turbid due to the coagulation of the meat proteins. Filter, and make the clear fluid up to the original volume by adding distilled water. Nitrogenous material in the form of commercial peptone is added in the proportion 1-2%. Add sodium chloride in the proportion 0·5% w/v. Dissolve by heating and filter.

Adjust the pH to 7·2 to 7·4 using NaOH or HCl as necessary taking care not to alter the volume very much. *Sterilise:* Autoclave 15lb/15 minutes.

Litmus milk: 2·5% by volume of alcoholic litmus solution is added to skimmed milk. This is sterilised by Tyndallisation.

Other liquid media: Media can be improvised from every day sources using materials which are readily available from the school kitchen. Any food which is suitable for human beings is also suitable for the growth of micro-organisms. Remember that:

It is easier to demonstrate microbial growth if the medium is clear to begin with and turns cloudy due to the increased number of organisms,

Broths should be free of fat.

These improvised media can be solidified using agar. They have the disadvantage that they are completely undefined and so are difficult to reproduce exactly. In order to repeat experiments and to compare results improvised methods should be standardised as far as is possible, making certain that each time the pH is·the same.

Solid Media

Nutrient gelatin: Edible gelatin is added to liquid medium in a concentration of 10-20% to convert it to a gel form. 15% is often used.

Sterilise. Use a high temperature for a short time, otherwise the gelatin will be denatured and will not gel on cooling. So sterilise in free steam for 10 minutes and then at 10lb/10 minutes. Store in screw cap bottles in which it should keep for a long time provided that the caps are properly screwed on. It is probably best made in 10 or 15 cm^3 quantities in test tubes or bottles. If you intend to make gelatin plates then larger quantities will be required.

Use. Since certain organisms possess enzymes capable of liquefying gelatine this property can be used to help identify organisms. Nutrient gelatin will support the growth of many organisms. (See page 29)

Nutrient agar. (See page 30) Agar is added to nutrient medium in the proportion of 2% in order to make nutrient agar. The whole is then placed in a steamer at 100°C for one hour to bring the agar to solution. Clearing the solution of particulate impurities and precipitated phosphates is done by filtration through filter paper or through cellulose wadding. Use a glass funnel (about 10 inches diameter) and fill the bottom third with glass beads or pebbles to act as a platform for the wadding. Make a thick pad of wadding, about two inches thick, moisten it with very hot water to ensure a good fit round the edges.

Keep the whole in a hot moist place, a steamer, until you are ready to use it. After filtering adjust the pH to 7·2—7·4.

Dispense into flasks or screw cap bottles and sterilise:

steam at 100°C/90 minutes, or
steam at 100°C/30 minutes on three successive days, or
autoclave at 109°C i.e. 5lb/30 minutes, or
autoclave at 121°C i.e. 15lb/15 minutes.

Dehydrated culture media

Granules: Dissolve the granules in distilled water to form a dense layer at the bottom of the container. Mix thoroughly, otherwise there may be deterioration and destruction of the medium during sterilisation due to superheating, hydrolysis, pH drift and so on.

Soak granules containing agar in distilled water for 15 minutes to allow them to swell and imbibe water. The granules will go into solution when brought to the boil. If you heat straight away a film of melted agar will form round the granules and act as an insulating layer against further heat penetration. The agar in the granules will dissolve on boiling and also, if the medium is being prepared in bulk prior to distribution to smaller bottles or flasks boiling ensures even distribution of ingredients. Failing to boil results in uneven distribution of constituents and a layer of undissolved agar settling at the bottom of the container. During boiling keep the mixture well stirred to prevent charring at the bottom. Use a sufficiently large container to allow for expansion and frothing. Cool to 50-60°C before you distribute it to its final containers, otherwise the hot medium will crack them. Mix the medium well before pouring.

Tablets. Usually simple to use e.g. one table to 10 cm³ distilled water into a test tube or bottle and heat sterilise. Refer to the Table 1 (page 20) for quantities and sizes of containers. Do not use too narrow a container because narrow tubes interfere with the self mixing action of this type of media.

Soak the tablets in cold water for 15 minutes. If you omit to do this a film of jelly will form round the tablets. Soaking causes them to disintegrate and they become self mixing. You can then dissolve them either by boiling or by autoclaving (see Fig. 8).

Oxoid tablets for agar media require: 1 tablet to 5 cm³ water. Thus:

plates: 2 or 3 tablets and 10-15 cm³ water
stabs: 2 tablets and 10 cm³ water
slopes: 1 tablet and 5 cm³ water.

Special cases: Some media in tablet form need special care. *Diagnostic sensitivity test agar* contains dextrose and a phosphate

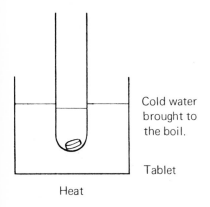

Cold water brought to the boil.

Tablet

Heat

Fig. 8

To show how tablets containing agar are dissolved.

buffer which may darken if overheated.

Media containing low amounts of agar: It is very important that the agar is well distributed by thorough mixing and by complete solution of the tablets. The weak gel may separate out afterwards if subjected to sudden cooling. Therefore it is important to allow the gel to cool slowly.

Ready made media

Oxoid Ltd. sell many media ready to use and will supply a list on request. For media which are required in small quantities or are difficult to make this a very convenient method.

General comments on sterilisation of media

Sterilise by boiling or by autoclaving. Do not sterilise in quantities greater than one litre because heat penetration becomes poor in large amounts.

Use containers which are large enough to provide a generous head for frothing.

Undo screw caps half a turn to allow a flow of tidal air during heat treatment. (Do the cap up fully and then undo it half a turn.) Screw down the caps after sterilisation is complete.

Containers should be alkali free to prevent alteration of pH. It is all right to make media using tap water if you are going to adjust the pH before sterilisation. By using distilled water with tablets or granules the correct pH will be attained.

Allow the steriliser to cool down to 60-65°C after sterilisation procedures before you open it otherwise the change in pressure in the container due to contraction of vapour inside will draw air in, and possibly also contaminating organisms.

Additions to the medium after sterilisation

Sometimes it is necessary to add substances to media after sterilisation procedures, either for convenience or because the substance may be denatured by sterilisation. Two common additives are blood and antibiotics. Care must be taken to observe aseptic precautions. If there is a large volume to be added that volume must be subtracted from the volume of fluid required to make or to reconstitute the medium. (Plan ahead.) When adding blood use fresh blood (human or animal, defibrinated) which has not been stored below 4°C. Warm up the blood in a water bath up to 37°C before adding it to agar which has been liquefied and cooled to 45-50°C. The volume of the container should be two and a half times that of the total liquid volume to allow for adequate aeration of the blood. (See page 30)

Remelting prepared media

All solid media will stand being remelted once after sterilisation and gelling. This means that media can be made, stored and used as required. Holding in the molten state or repeated melting and gelling will destroy media.

Storage of media

Dehydrated media keep at an even temperature in a cool dry place away from direct light. They are hygroscopic so that once they are opened they must be protected from atmospheric moisture, and the inner seal of the bottle and the lid replaced. Do not open the bottle in a steamy or humid atmosphere.

Reconstituted media: These can be stored in a refrigerator at 4°C or in a cool dark place such as a cupboard. They will keep better in screw capped bottles rather than in ones with cotton wool or other type of stopper. It is best to pour solid media at as low a temperature as is practicable (see later) in order to allow excess moisture to evaporate before refrigeration and to minimise condensation water.

Prepared media: Store at 4°C in a cool dark place. Humidity is not an important factor as the containers are sealed.

3. Uses of media

Liquid media for bacteria

Nutrient broth or meat infusion broth (pH 7·4 approx.) (See page 33)

This is used to encourage the growth of organisms. It is a good general purpose liquid medium, and is used as the basis for nutrient agar.

Peptone water (pH 7·2 approx.) (See page 23): This is used as *growth* medium and as the basis for sugar media used in *sugar fermentation tests.* The sugars most commonly employed are given as a table on page 23 .The sugar or fermentable substance is added to peptone water in the proportion of 0·5 or 1%, and an indicator added to detect acid change:

Neutral red (0·25% of a 1% solution)—if acid is present it changes to pink.

Phenol red (0·01%) if acid is present it changes to yellow.

Gas production is detected by the addition of a Durhams tube to the medium.

Peptone water is also used to detect the presence of *indole*—some organisms can utilise the amino acid tryptophane and .produce indole—a test used in differentiating coliforms. To detect indole add to a 48 hour culture Ehlichs rosindole reagent—a positive reaction is pink.

Litmus milk: This is used to test the saccharolytic (fermentation of lactose) or proteolytic (digestion and coagulation of casein) properties of some organisms; and for growing dairy organisms (lactic acid organisms.) It shows the following reactions:

reduction of litmus—the medium becomes colourless

acid production—the medium becomes pink

alkali production—the medium becomes blue

acid .clot—a soft gelatinous clot which does not retract and is completely soluble in alkali

rennet clot—a firm clot which eventually retracts and expresses clear greyish fluid, and is not soluble in alkali,

'stormy clot'—the clot which is broken up by acid production. (See page 24)

Cooked meat medium: a peptone, beef extract and heart tissue medium. It can be used for aerobes and anaerobes, but is particularly useful for the latter. (See page 59)

For the growth of anaerobic organisms it is best used sealed with a layer of sterile paraffin wax. *Clostridia*—saccharolytic species produce acid and gas, the medium smells and meat may redden; proteolytic species—the medium becomes foul smelling, proteins are broken down and the meat may blacken.

Solid media for bacteria

Nutrient gelatin Stab cultures to detect the ability to liquefy gelatin. (See page 25)

Nutrient agar. (pH 7·2 approx.) Very widely used as a general purpose medium and is indispensable. (See page 25)

MacConkey's bile salt neutral red lactose agar (and broth) (pH 7·4) This is a peptone solution solidified with agar and to which has been added 0·5% bile salt, 1·0% lactose and an indicator—either neutral red (pH 6·8 red, pH 8·0 yellow) or bromo-cresol purple (pH 5·2 yellow, pH 6·8 purple). Organisms which have fermented lactose to lactic acid can be distinguished from non lactose fermenting organisms by local changes of the indicator around the colonies. Thus with neutral red—non lactose fermentors, pale and lactose fermentors acid—pink bromo-cresol purple—non lactose fermentors, no change, lactose fermentors, yellow. This medium is generally used for differentiating intestinal organisms of the coliform group (which ferment lactose) from related pathogens such as Salmonellae which do not. It is used in the examination of water, milk, ice-cream.

Desoxycholate citrate agar (pH 7·0 approx.) This meat extract and peptone medium contains lactose, sodium and ferric citrate, sodium thiosulphate, sodium desoxycholate, and neutral red as an indicator. It is used in the identification of organisms of the Genus Salmonella (food poisoning strains, typhoid, paratyphoid) and the Genus Shigella (dysentery). These organisms are non lactose fermenting and can be detected as pale colonies on the medium—other non lactose fermenting non-enteric organisms are inhibited by this medium.

Milk agar: This is a medium used for plate counts on milks, rinse waters, ice-cream and so on. 2 volumes 3% nutrient agar are mixed when melted and cooled to 56°C with 1 volume of commercial sterilised milk. It is not a selective medium.

Blood agar (See also page 27) Nutrient agar is made, sterilised and cooled to a temperature (about 50°C) very slightly above its gelling point. Sterile human or horse blood (defibrinated) is added in the proportion of 10-15% by volume using a sterile pipette. The blood is quickly mixed by rolling the tube or container between the hands. The blood agar is poured as a plate or is allowed to set as a slope. This enriched medium is used for the culture of fairly delicate, nutritionally exacting pathogens, and other commensal organisms for example, Streptococci, Haemophilus. It is also used to demonstrate the haemolytic ability of an organism: haemolysins diffuse into the agar around the colony, lyse red blood cells and a zone of clearing is produced—some Staphylococci and Streptococci have this ability.

Tomato juice agar (pH 6·1 approx.) This medium contains tomato juice solids, peptone, peptonised milk and agar. Because of the growth factors provided by tomato juice solids it is a good medium for the growth of Lactobacilli from milk products and saliva.

Rogosa agar (pH 5·8 approx.) Trypone, dextrose, yeast extract and inorganic salts are the main constituents of this medium—and is another good medium for growing lactobacilli from dairy products. Lactobacilli tend to be anaerobic or microaerophilic and should be incubated in an atmosphere of 95% hydrogen : 5% carbon dioxide. If this is not possible, overlay the plate with another layer of tempered Rogosa medium after inoculation.

Tryptone soya agar (pH 7·3 approx.) A good nutrient medium for the growth of a wide range of organisms.

Tributyrin agar (pH 7·5 approx.) A peptone and yeast extract medium containing tributyrin, and which is opaque when made. It is used to identify and enumerate lipolytic (fat splitting) organisms e.g. from butter. Lipolytic activity is indicated by a clear zone around the colony due to the diffusion and action of lipases, hydrolysing the tributyrin.

A medium for the growth of marine organisms. (Ref. 3) Filter a sample of seawater. Add 0·2% Casitone (Difco) 0·1% yeast extract. Adjust the pH to 7·5 and then boil. Filter the solution, dispense it into its final containers and then autoclave it in the normal way. This makes a broth. For plates and slopes add 1·5% agar. It can be buffered with Tris* in 0·05% concentration.

Media for fungi

Liquid

Sabouraud's glucose broth (pH 5·4 approx.)
Malt extract broth (pH 5·4 approx.)

Solid

Sabouraud's glucose agar (and broth) (pH 5·4 approx.) This medium contains 4% w/v glucose and 1% w/v mycological peptone and is a good medium for the cultivation of moulds, yeasts and acidophilic bacteria.

Malt agar (and broth) (pH 5·4 approx.) This medium contains 3% w/v malt extract and about 0·5% w/v peptone and again is good for the growth of yeasts and moulds.

*Tris buffer = Tris (hydroxymethyl) amino methane

Czapek Dox medium (pH 6·8 approx.) is a completely defined medium in which sodium nitrate (0·2%) w/v is the sole source of nitrogen. It is a good medium for the cultivation of moulds and yeasts.

All these media can be purchased in the form of granules which obviates the need to make them from basic constituents.

4. Manipulation
Manipulation of Liquid Media

Liquid media are usually dispensed into their final containers before sterilisation so that any extra unnecessary manipulations after sterilisation are cut out.

The quantities of media which can safely be placed in containers has been dealt with before in Table 1 (page 20)

The usual quantities of liquid media for cultures are listed as in Table 4.

Table 4 Quantities of liquid media

1. Peptone water for biochemical tests	5 cm^3 in a test tube
2. Peptone water sugars for sugar fermentation tests	5 cm^3 in a test tube
3. Nutrient broth or other broths for cultures	10 cm^3 in a test tube or screw cap bottle
4. Cooked meat media for culture of anaerobes	20 cm^3 in a screw cap bottle

Durham's tubes are filled as follows: the liquid medium is put into the test tube, and the Durham's tube dropped into it with the concavity uppermost. Sterilisation procedures will cause the air at the top of the tube to be expelled and on cooling the space will be completely filled with the medium. (See Fig. 9)

Manipulation of Solid Media

Stabs: Deep media are used. The solidified medium is dispensed into test tubes or bottles, sterilised and allowed to set while the tube is in a vertical position. The volume of medium used is 15 cm^3. Stabs of *gelatin* are made in test tubes (because bottles with rubber seals must be autoclaved and gelatin will not stand that), and they are used for tests of the ability of organisms to liquify gelation. Stabs of N.A.* are made in either test tubes or bottles. This is a culture method for the growth of anaerobic organisms.

Slopes: N.A. and N. gelatin are used. The medium is made and sterilised in a suitable test tube or screw cap bottle. After sterilisation the container is placed in such a position that it sets forming a very large flat surface. It is usual to place the tubes or

*Throughout the text N.A. indicates nutrient agar

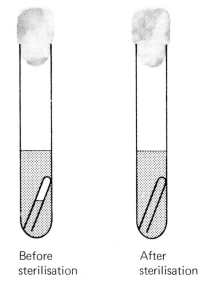

Before
sterilisation

After
sterilisation

Fig. 9
How Durham's tubes are filled

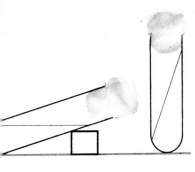

10

w slopes are set and made

bottles so that they are nearly horizontal, and with their necks supported on a glass rod. They should be left undisturbed until the medium is quite solid. When making slopes it is best to wait some time before the sterile media are removed from the steriliser. Agar containing media will stay liquid down to about 50°C. If the media are not subjected to rapid changes in temperature the quantity of condensation water will be minimised which is very desirable. (See Fig. 10)

Table 5 Quantities of media for slopes:

2-3 cm³ in a bijou bottle
10 cm³ in a universal bottle
5 or 10 cm³ in a test tube.

Plates: A petri dish containing solid nutrient medium set in a layer in the base is known as a **plate**. The quantity of medium used must be sufficient to form a layer which will completely fill the base. 12—15 cm³ is a satisfactory quantity, 10 cm³ will suffice but great care (and skill) is needed to make a complete agar layer before the coolness of the petri dish prematurely sets the agar.

How to pour a plate. (assuming right handedness . . . reverse the hands for left handedness)

Prepare the bench. Have a lighted bunsen burner with the air hole open, a test tube rack, and the sterile petri dish.

Take the sterile petri dish and, if necessary, remove it from its brown paper wrapping taking care not to open the lid. Place it on the bench near your *left* hand, lid uppermost, but still shut.

In the meanwhile a test tube (or bottle) containing 12-15 cm³ of sterile N.A. which has been melted in a boiling water bath, and has now been cooled down to a temperature slightly above its gelling point. Keep it at this temperature until it is required. (Remember that once that it has been allowed to gel it must be reheated to about 100°C to melt again).

Fetch the tube of melted agar and *quickly* do the following:

Holding the tube in your right hand, with the third and fourth fingers and the side of your left hand remove the stopper and place it on the bench. Flame (see page 39) the top of the tube. With your left hand lift the right hand side of the lid and tip the melted agar into the base. Replace the lid of the petri dish and put the tube down. With your right hand gently rotate the base so that the agar forms a complete layer. This series of actions should take only a few seconds. Leave the freshly poured plate flat on the bench for some minutes while the medium forms a firm gel. Do not disturb it.

If when you pour the plate some air bubbles are caught in the surface of the agar you can get rid of them by cautiously *waving* the flame of the bunsen burner over the surface of the still liquid medium. Care must be taken not to over do this otherwise you may crack the glass petri dish or destroy a polystyrene one.

You will notice that in freshly poured plates that a varying amount of condensation water forms on the lid. The plate must be dried to rid it of this water which otherwise might drop onto the surface of the medium and interfere with culture growth.

Drying the surface of agar plates: Take the plate and lid to the 37°C incubator. Open the incubator door. Then cautiously open the lid and base of the plate in such a way that throughout the procedure the inside of BOTH parts face downwards. This lowers the· risk of any airborne germ alighting on the sterile surfaces. Then place the lid, inside down, on the shelf and prop the medium-containing base also inside down against it so that it rests at an angle to the shelf. Close the door of the incubator. (See Fig. 11).

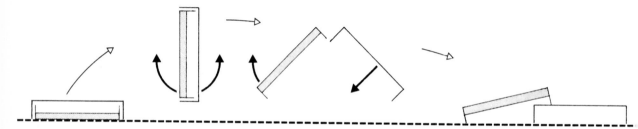

Fig. 11
Stages in opening a plate in
order to dry it

Leave the two halves of the plate like this for 10 to 15 minutes or until the condensation has evaporated. The lid and base must be paired before removal from the incubator, again observing aseptic precautions. If many plates are to be dried they must be stacked in the incubator with alternate lids and bases resting against each other. If this is done great care must be taken not to contaminate the medium of one with the base of another. (See Fig. 12).

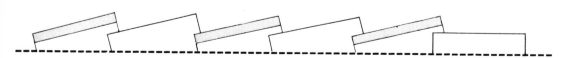

Fig. 12
Plates drying in the incubator

For some experiments it is essential that the surface of solid media in plates is very dry. In these circumstances the plates are left in the incubator as described for two hours.

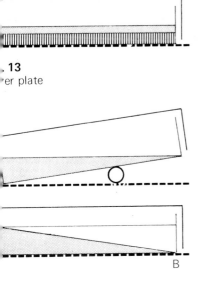

13
er plate

14
ges in making a slope plate

Layer plates: These are plates in which layers of different media are poured into the same plate. e.g. *blood plates.* In order to economise on blood the following technique can be adopted: pour 10 cm³ N.A. onto the bottom of a plate and allow it to set. Pour 10 cm³ of cooled N.A. mixed with 1-2 cm³ defibrinated horse blood on top and allow to set. (See Fig. 13).

Slope plates: (gradient plate.) These are layer plates in which the layers are at an angle to one another. e.g. disinfectant plates. 10 cm³ of N.A. containing a disinfectant are poured into a petri dish so that the disinfectant layer sets at an angle. (See Fig. 14(a)).

When set 10 cm³ N.A. is poured on top so that the surface is flat. (See Fig. 14(b)). Once the disinfectant has diffused between the two layers there will be a concentration gradient across the plate from *A* to *B*.

Prevention of the dehydration of media: Since cultures are not normally incubated for great lengths of time in normal circumstances, the media do not dry out. However on rare occasions dehydration may be a problem.

(i) Using a large tin container. After a week a medium may begin to dry out. This can be counteracted by incubating plates inside a large tin (a 3kg biscuit tin for example) and which contains an open jar of water which will evaporate and keep the atmosphere moist. It is a possible difficulty with culturing slow growing moulds.

(ii) Use larger quantities of media—this will delay bad dehydration.

Stock cultures will not suffer dehydration provided that they are set up in screw cap bottles.

Storage of poured sterile media: it is not a good idea (from the asepsis point of view) to attempt to store poured plates, unless of course they have been purchased and are sealed in their polythene wrappers. It is much sounder policy to keep stocks of sterile medium in containers which have a low risk of accidental contamination. The medium can be made or reconstituted in a suitably large vessel but *before* it is finally sterilised it should be distributed into vessels suitable for its storage. It is a good idea to store media in small quantities, that is—in the exact volumes in which it will be required:

plates 10-15 cm³
stabs   10 cm³
slopes   5 cm³

The above media can be both sterilised and stored in screw cap bottles. This has the added advantage that if any contamination does occur it will be limited to a small amount of medium. Media stored in tightly screwed up bottles are not subject to loss by evaporation. They are stored in a cool dark cupboard or at 4°C in a refrigerator.

3 Sterilisation

Aseptic technique

Aseptic techniques are methods which are adopted in micro-biology for two reasons: to prevent organisms getting into cultures and forming mixed cultures, and also to prevent culture organisms, which are possibly pathogenic, being distributed into the laboratory environment.

For most experimental purposes organisms are grown in pure culture. This means that any observed changes can be attributed to the species of organism present. Unwanted organisms, forming mixed culture will also produce changes which may lead to the wrong interpretation of experimental results. So it is vital that pure cultures are kept pure. Aseptic methods allow for the transfer of chosen organisms from one medium to another but obviate the risk of contamination.

When cultures are set up it is vital that all equipment glassware and media which are to be in direct contact with the culture are sterile. The following sterilisation methods are designed so that the organisms present on equipment and so on are killed and that these items remain sterile right up to the moment of use. The time lag between sterilisation and use of equipment should be of no importance provided that an aseptic technique is adopted. Similarly the methods used in growing and maintaining cultures are designed to minimise the risk of contamination by extraneous organisms. Avoidance of labels other than self adhesive ones; never eating, smoking, or sucking the tops of pens and pencils; indeed any practice which minimises risk of contamination contributes to asepsis.

Principles of sterilisation

The principle of sterilisation is to treat equipment and media in such a way that after treatment, there is no vegetative organism or resistant spore form left alive in, or on, that piece of equipment. Items so treated will then be sterile.

In sterilisation procedures certain criteria must be met. There are two forms of micro-organisms—the vegetative forms which are relatively easy to kill and the resistant or spore forms which are much more difficult to kill. Therefore in general the aim is to treat equipment in such a way that if the most resistant type of spore known were present it would be killed. A successful sterilisation technique will achieve this. So:

All methods of sterilisation must kill or remove all forms of life present.

The treatment must not alter the chemical or physical composition of media (note sugars, gelatin).

Media must be sterilised in such a way that loss of water by evaporation does not take place.

Allowance must be made for the quality of the materials from which equipment is made so that, for example, heat treatment does not crack glassware, or destroy the rubber seals of bottles.

Lastly, things which have been sterilised must remain sterilised.

There are three methods of sterilisation commonly used by micro-biologists:

Heat

Chemicals

Filtration . . . this is not a practical method for schools.

Heat Sterilisation

This is widely used, and different forms of heat are employed which are summarised in Table 6.

Table 6

Dry Heat	Moist Heat

1. Flame   red heat
              flaming
2. Dry hot air
      160°C/60 mins.
	1. Boiling water. 100°C
	2. Saturated steam in a steamer 100°C
	3. Autoclave 105-135°C. To kill the most resistant spores needs moist heat at 121°C for 15-30 mins.

Living cells and spores are killed because of destructive oxidation of the cell constituents	Living cells and spores are killed because of the coagulation of structural proteins and essential enzymes.

Whichever method of heat sterilisation is most suitable for the job in hand there are several factors influencing the killing of microbes which must be considered:

temperature

exposure time

presence or absence of moisture

the form of the organism: vegetative or spore

the nature of the infected material

It has been shown that temperature and time are inversely related—the greater the temperature the less time is required for sterilisation, and that the presence of moisture speeds up sterilisation processes considerably. For example to be sure of achieving sterility using dry heat the normal exposure is for one hour at 160°C; using moist heat 15 minutes at 121°C.

It is important to realise that these times do *not* include the heating up time.

The spores of Actinomycetes, yeasts and fungi are more resistant than their vegetative forms but less so than bacterial spores.

Thus to summarise:

Most susceptible vegetative cells are killed 70°C for 5 mins.

Most resistant vegetative cells are killed 80-90°C for 30 mins.

Most bacterial spores are killed in the range 100-121°C for 10 mins.

Red Heat

Equipment: Bunsen burner.

Some pieces of equipment can be sterilised by heating them in a bunsen burner flame (air hole open) until they become red hot.

Applications: This is used mainly for sterilising wire loops and straight wires.

Method: Hold the handle of the wire loop at an angle so that the whole of the bottom 2/3 of the wire are in the hot part of the flame *at the same time*. Wait until the wire becomes red hot, that is retains red heat for 15-30 seconds; remove from the flame and allow it to cool. The bottom 2/3 of the wire is now sterile. Rest the sterile loop in a position where it will not come into contact with anything and will therefore not become recontaminated. It is all right to use the loop immediately or shortly afterwards without re-sterilising, but if you are in *any* doubt as to the sterility of the loop resterilise it. It is worth the bother. (See Fig. 15.)

Flaming

Equipment: Bunsen burner.

Flaming comprises heating the surface of equipment with a bunsen burner flame to kill any surface contaminating organisms, in cases where red heat would be damaging.

Applications: The mouths of test tubes and flasks are flamed when they are opened and both before anything is put into and after anything is drawn out of them.

Method: Remove the lid or stopper and pass the mouth and 1½ inches of the top of the tube or flask *through* the hot part of the flame two or three times. Carry out subsequent operations then prior to the replacement of the lid or stopper reflame the neck. This is to kill any organisms which may have accidentally got onto the top. Care must be taken not to overheat otherwise the glass will be cracked.

In the field, for example, when collecting samples of river water the same procedure is used. The mouth of the sample bottle is

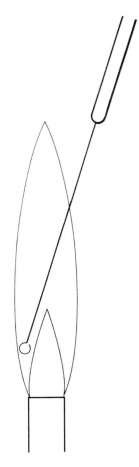

ttom 2/3 of the wire in the hot
rt of the flame

g. 15
oop being sterilised

(a)

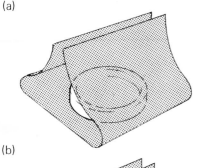

(b)

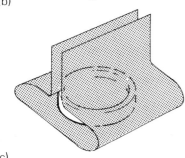

(c)

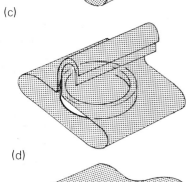

(d)

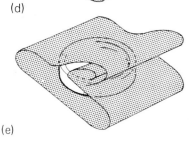

(e)

Then fold in the
ends—like wrapping
a parcel

Fig. 16 Stages in wrapping a petri
dish prior to sterilisation

sterilised to free it from extraneous organisms prior to the collection of the water sample. The flame of a cigarette lighter or a small methylated spirits stove will substitute well for the bunsen burner.

Forceps, knives, scalpels, needles, and so on may be sterilised by flaming when the temper of the steel would be destroyed by red heat.

Hot air

In the laboratory the hot air oven is used as a piece of sterilising equipment. Sterilising ovens are usually heated by electricity and the heat is thermostatically controlled.

Applications: Dry hot air is the best method for sterilising dry glassware such as test tubes, flasks, petri dishes, pipettes, all glass syringes.

Sterilisation temperature: A holding temperature of 160°C for 1 hour is usually used. This does not include heating up time and is timed from the moment the holding temperature 160°C is reached.

Preparation of equipment for sterilisation with hot air: All equipment should be dry otherwise there will be a tendency for it to crack when heated.

(a) Petri dishes: These should be clean and dry and each base paired with its lid. They may either be packed into sterilising cans or they should be wrapped individually in Kraft paper (brown paper). (See Fig. 16.) When sterilisation has been completed each petri dish so wrapped will remain sterile until it is unwrapped to be used. There is a big risk that dishes which are not wrapped or placed in tins during sterilisation will very soon become contaminated after removal from the oven.

(b) Test tubes: Each tube should have either a cotton wool stopper or a metal cap. They are sterilised either in metal racks or in tins (cocoa tins are suitable). No lid for the tin is necessary in this case as it is merely to support the tubes, not to maintain their sterility.

(c) Flasks: Each flask should have either a cotton wool stopper or a metal cap. The flasks should be empty and dry. Flasks of media should *not* be sterilised in the hot air oven (See Autoclave p. 42).

(d) Pipettes: These should be plugged with cotton wool (See Chapter 1.) Drawn or undrawn pasteur pipettes are sterilised in bundles inside closed tins. Other pipettes are wrapped individually in long thin strips of Kraft paper. The ends of the Kraft paper are twisted so that the paper strip does not untwine. After sterilisation they remain wrapped until just prior to use.

(e) Screw cap bottles: The rubber liners of the caps will not withstand sterilisation temperatures and bottles and their caps are sterilised in the autoclave.

Procedure: The oven must not be packed too full otherwise the circulation of air will be impeded. Load the oven, shut the door and then allow heat up to sterilisation temperatures over 1 or 2 hours, leave it at 160°C for one hour, turn it off and then, before the door is opened, allow it to cool down. Glassware subjected to sudden changes in temperature may crack. Do not use the hot air oven for melting media because the water in the medium will evaporate.

Boiling water bath

This may be merely a beaker of water heated over a Bunsen burner, or may be something more permanent for example a metal container, an old saucepan or a fish kettle over a gas ring, all of these are suitable. The bottom of the bath should contain a perforated tray so that the contents of the water bath are not in direct contact with its base, will not be subject to bumping when the water boils and will be easily lifted out. It is best to use soft or distilled water because hard water may leave a deposit on glassware.

Sterilisation is promoted by adding sodium carbonate in a concentration of 2%.

Applications: For sterilising tubing, pipettes, measuring cylinders, rubber stoppers, scalpels, forceps, scissors and other metal instruments which have been used and are known to be contaminated. For example, after a sterile pipette has been used to transfer a quantity of live culture from one tube to another the pipette can be placed directly in a water bath so that the remaining viable organisms can be killed. After this sterilisation the pipettes are then taken, dried, wrapped and re-sterilised in the hot air oven. Measuring cylinders: care must be taken that the base does not crack off at the seal.

Sterilising time: 100°C for 5-10 minutes. It is *not* adequate for killing resistant spores.

Note: A boiling water bath is a good method for melting media provided that the depth of the water is higher than the level of the medium in the container. (Refer to Fig. 8)

Steaming at 100°C

Equipment which cannot desirably be immersed in water can be

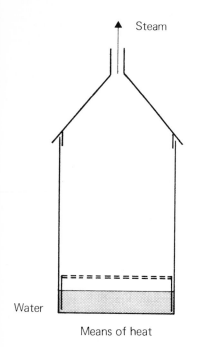

Steam

Water

Means of heat

Tubes and bottles of media are placed on perforated shelf.
(cover cotton wool stoppers with Kraft paper.)

Fig. 17
A steamer

sterilised in free steam at 100°C, because steam has good penetrating power. A steamer consists of a deep cylindrical metal container which has a close fitting lid with a steam outlet. It contains water in the bottom which is heated from beneath to produce steam. There is a perforated shelf in the bottom above the level of the boiling water. (See Fig. 17)

Applications: Media can be satisfactorily sterilised in free steam. It is also good for melting media.

Method: Items are packed loosely on the shelf, the lid is placed in position and the water is heated to boiling. Steam will eventually issue from the steam outlet. For broths and other nutrient media the sterilisation time is a single exposure at 100°C for 90 minutes. This time includes the heating up time, but for large volumes it is advisable to sterilise longer to allow for heat penetration.

Table 7 Heat penetration times:

up to 100 cm^3	15-20 minutes (Total sterilisation time = 90 minutes)
up to 600 cm^3	30 minutes Add 15 to 30 minutes to the total time.
up to 5 litres	45 minutes (Total sterilisation time = 90 mins. PLUS 15-30 minutes.)

For delicate media which might be damaged by long sterilisation times sterilise by the method known as Tyndallisation. This is an intermittent process which involves exposure to steam for 20-45 minutes on 3 successive days. Use this method for fermentation sugars, and for gelatin containing media (which otherwise loses its gelling properties). Tyndallisation kills vegetative forms on first exposure. Overnight any spores present will germinate and be killed on the second exposure. The third exposure will ensure that any spores which escaped and subsequently germinated are killed.

Autoclave

Saturated steam is applied at temperatures above 100°C. Water is boiled in a closed vessel under pressure and therefore the temperature reaches greater values than 100°C, dependent on the pressure. For most purposes the autoclave is set at 15 lb above atmospheric pressure. If no air is present in the autoclave the temperature will be 121°C. Large autoclaves, such as are used in microbiological laboratories are expensive and more than likely above the means of a school budget. (If the budget can afford it I would strongly recommend buying one because a good autoclave makes life much easier.) Many manufacturers sell what are called portable autoclaves. These are relatively cheap and are very good.

Alternatively a fairly large domestic pressure cooker will hold a large number of universal bottles (4 dozen or so.) or two piles of six petri dishes. Small flasks containing media will fit in also.

Applications: Autoclaves are used for sterilising all media which will withstand severe heat treatment. They are also used for killing cultures which are no longer required.

For most purposes a pressure of 15 lb for 15 minutes will sterilise an article.

Method for using a simple autoclave

1. See that the autoclave contains enough water (3½ inches for a 19 inch autoclave).

2. Put in the items to be sterilised and turn on the heater.

3. Place the lid in position.

4. Open discharge cap.

5. Screw the lid down by sequentially tightening opposite pairs of handles.

6. Adjust the safety valve to the desired pressure. (See Table 9)

7. Boil water until steam discharges and is expelling the contained air. Allow this to happen until only steam is issuing. This can be checked by leading the steam into a bucket of water . . . no bubbles means no air. Remove rubber tube from water. Then close the discharge cap.

8. Pressure now rises until steam issues from the safety valve, when the internal pressure has reached the preset value.

9. From this point hold for the holding period, usually 15 minutes.

10. Turn off the heater and allow the pressure cooker or autoclave to cool until the pressure gauge indicates atmospheric pressure inside.

11. At once open the discharge tap and allow air to enter *slowly.*

Note: If opened too quickly liquid media will boil at the rapidly reduced pressure and may therefore spill inside the autoclave. If the internal pressure is allowed to fall much below atmospheric pressure excessive evaporation may take place from the media.

Preparation of items for sterilisation in the autoclave.

1. Flasks: Flasks containing media should have either a cotton wool stopper or a metal cap. In addition, a piece of Kraft paper should be securely tied with a piece of string (or a rubber band which will probably withstand the heat) over the top and round the neck of the flask. This prevents excessive condensation of water in the cotton wool, and is an additional safeguard against contamination if the media are to be stored before use. For the sterilisation (that is the disposal) of old cultures there is no need to cover the tops with Kraft paper.

2. Test tubes containing media should be stoppered either with cotton wool or with a metal cap, and a piece of paper laid over the tops for the same reason as above. This paper is disposed of after sterilisation. (Empty dry clean test tubes are sterilised in the hot air oven.)

3. Petri dishes containing old cultures are piled up with their lids uppermost (do not separate lids from bases). Do not try to pack the dishes in tightly. It is better to have two sterilisation sessions.

4. Screw cap bottles: Screw the lid of each bottle up to its fullest extent and then unscrew it half a turn. This will loosen the cap sufficiently for safety during and after sterilisation. Both dry empty, and bottles containing media are autoclaved. After sterilisation the lids are tightened up.

Table 8

Container	Size	Quantity it can safely hold during sterilisation
Test tube rimless	5 inches x ½ inch	up to 4 cm^3 medium
	6 inches x 5/8 inches	up to 5-10 cm^3 medium
	6 inches x ¾ inch	up to 10-15 cm^3 medium
	7 inches x 1 inch	up to 20 cm^3 medium
		Greater than 20 cm^3 use a flask or bottle.
Flask	250 cm^3	up to 100 cm^3
	1 litre	up to 400 cm^3
	2½ litre	up to 1 litre
Bottle	24 cm^3	up to 10 cm^3
	8 cm^3	up to 2-3 cm^3

Disinfection using chemicals

The other method, open to the teacher, by which micro-organisms may be destroyed but only applicable in certain circumstances is the use of disinfectants. Sterilisation by filtration is not practicable in schools.

Many substances have disinfectant properties and the choice of substance depends on certain factors:

1. As a general rule disinfectants cannot be relied on to sterilise—most bacterial spores can withstand disinfectants.

2. Concentration of disinfectant: the more dilute an agent the less effective it will be. Use of very dilute disinfectants encourages the selection and survival of disinfectant resistant cells and may permit their growth and multiplication.

3. Organic matter such as food waste, faeces, grease etc., greatly

reduce the effectiveness of disinfectants and they should there-fore, when practicable be removed before the disinfectant is used.

4. A solution of disinfectant becomes less effective the older and more used it is. Pots of disinfectant should be changed regularly and frequently. This is important.

5. Up to about ten minutes the longer the time of exposure the more effective the disinfection process. Beyond ten minutes time of exposure is less important, and the concentration more important.

6. Increased temperature aids disinfection.

7. The nature of the substrate on which a disinfectant is applied must be considered—some substances can be damaged.

8. Many disinfectants, especially when concentrated can be damaging to the skin—care should be taken in their use.

Applications of disinfectants in the laboratory

1. Provide 2-3 jam pots of disinfectant per bench of pupils into which used slides can be disposed.

2. Tall jars (preferably of plastic or polythene) containing disinfectant can be used to receive used pipettes.

3. Wash bottles of disinfectant can be used to 'spray' down the benches after every practical period in which micro-biology is a part.

4. A bucket of disinfectant into which used cultures—plates and tubes—can be disposed is very useful. This should not be the sole means for their disposal but is a preliminary step before heat sterilisation. Use wooden tongs (such as are used to remove washing from a washing machine) to remove items from the bucket. The tongs should be stored in disinfectant.

5. Have jam pots/wash bottles of disinfectant ready to use in case any pupil drops culture on the bench or floor or on themselves. Cover the affected area liberally with the agent and leave it for five minutes before clearing it up.

Concentration of agent to use

It is not possible to generalise on the concentration of agent to use because this depends on the nature of the agent and the purpose for which it is to be used. For general purposes lysol is used at about 3% solution, hypochlorites diluted to give about 200 p.p.m. available chlorine. Tests can be done to ascertain effective working concentrations—see Chapter 7 — Antimicrobial substances.

Some antimicrobial agents

The following groups of substances are antimicrobial: acids, alkalis, phenols, alcohols, aldehydes, ketones, solutions of salts of heavy metals, organo metallic compounds, hypochlorites, quaternary ammonium compounds.

Avoidance of Sterilisation

It is not possible to totally avoid sterilisation but it is possible to avoid some of the more laborious sterilisation procedures. Media and petri dishes can be bought sterilised and ready for use. As mentioned before, this type of petri dish can only be used once and must then be discarded. Discarded cultures can be disposed of by burning the whole in the school incinerators or furnaces.

Plates can be bought ready sterilised and poured.

Maintenance of Sterility

Items once sterilised must be kept sterilised. Micro-organisms are ubiquitous. Lids of flasks, tubes, plates, time and any other containers must be kept shut except when really necessary, and then only opened for the minimum length of time. Wrappings must remain sealed or closed until the moment before use. Loops and wires must be sterilised the *moment* before use and again the moment *after* use. *It is not possible to be too careful.*
Remember if you are in any doubt *resterilise.*

Table 9 Summary of sterilisation methods.

Bottles, screw cap, with media	Autoclave	121°C/15 minutes
Bottles, screw cap, clean	Autoclave	121°C/15 minutes
Cultures, old	Autoclave	121°C/15 minutes
	Steamer	100°C/90 minutes
	Disinfectant	
Flask, dry, clean.	Hot air	160°C/90 minutes
Flask, mouth	Flaming	Few seconds
Forceps	Flaming	Few seconds
Forceps, contaminated	Boiling water	100°C/10-15 minutes
Gelatin	Steamer	100°C/90 minutes (Tyndallisation)
Knives	Flaming	Few seconds
Measuring cylinder	Boiling water	100°C/10-15 minutes
Media delicate	Steamer	100°C/90 minutes (Tyndallisation)
Media, N.A. and other	Autoclave	121°C/10-15 minutes
	Boiling water	100°C/10-15 minutes
	Steamer	100°C/90 minutes
Media, N.B.*	As N.A.	
Media, sugar containing	Steamer	100°C/90 minutes
Needles	Flaming	Few seconds
Petri dish, dry clean	Hot air	160°C/90 minutes
Petri dish, old culture	Autoclave	121°C/15 minutes
	Boiling water	100°C/10-15 minutes
Pipettes, contaminated	Boiling water	100°C/10-15 minutes

Pipettes, dry clean	Hot air	160°C/90 minutes
Rubber stoppers, contaminated	Boiling water	100°C/10-15 minutes
Saline, see Media		
Scalpel	Flaming	Few seconds
Scalpel, contaminated	Boiling water	100°C/10-15 minutes
Scissors, contaminated	Boiling water	100°C/10-15 minutes
Seeds	Dip in $HgCl_2$ or other heavy metallic salt solution to disinfect, wash afterwards in sterile distilled water.	
Soil	Steam it over water or saturate with formaldehyde and leave until no further smell.	
Spreader, glass	Boiling water	100°C/10-15 minutes
Sugar containing media	See 'Media' delicate	
Test tube, dry	Hot air	160°C/90 minutes
Test tube containing medium	See media	
Test tube, mouth	Flaming	Few seconds
Tubing, contaminated	Boiling water	100°C/10-15 minutes
	Autoclave	121°C/15 minutes
Vegetable matter	As seeds, or with ethylene oxide	
Water (distilled, saline)	See media	
Wire loop	Red heat	15-30 seconds
Wire, straight	Red heat	15-30 seconds

*Throughout the text N.B. indicates nutrient broth

4 Cultures

This chapter describes the techniques which enable the growth of selected micro-organisms and which obviate the risk of contamination by unwanted species, in both liquid and solid culture. It is divided into parts, the order of which follows the sequence of actions taken in growing micro-organisms. The manipulation of sterilised media, intimately associated with the growth of cultures was dealt with in the last section of chapter 2 on media. This chapter deals with the cultures themselves—how they are inoculated, maintained and incubated.

Assemble the following equipment:

Bunsen burner (alight with the air hole open)
Wire loop
Straight wire
Test tube rack
Tubes of sterile liquid or solid medium.
Cultures.

A. Inoculation

Inoculation means the introduction of micro-organisms (or sometimes, chemicals) into or onto a sterile medium.

There are four basic techniques:

Removal of organisms from liquid culture,
Removal of organisms growing in or on solid culture,
Inoculation of liquid media,
Inoculation of solid media.

Combinations of these four cover all the various methods of inoculating.

Removal of Organisms from Liquid Culture

(a) **Using a loop:** Sterilise loop by flaming, and allow it to cool. Hold the culture tube in the left hand (assuming right handedness) so that the bottom half of the tube lies against the closed fore and second fingers. Hold the tube in place with the thumb. Holding the loop in the right hand remove the cotton wool plug, or metal cap with the 3rd and 4th fingers of the right hand. Pass the top of the culture tube *through* the flame a couple of times and then insert the sterile loop into the medium. Do not dip it any deeper than is necessary to obtain a loopful, so that the film stretches across the loop. Then remove the loop from the tube avoiding touching the sides of the tube. Flame the top of the test tube and then replace the stopper.

Do not put the loop down anywhere and remember that it

contains living, and possibly pathogenic organisms.

With a screw cap bottle the technique is basically the same but manipulation of the screw cap can be awkward. It is best to loosen the lid before the loop is sterilised so that at the point when the lid is to be removed with the side of the right hand it does not present difficulties. The screw cap lid should NOT be put down during operations.

(b) Using a pipette. (A pipette is always used with a mouth piece.) Unwrap the mouth end of the pipette, leaving the operating end still covered. Do *not* remove the cotton wool plug when you attach the mouth piece. Carefully remove the rest of the wrapping taking care not to touch the pipette, but support it by its upper end. Hold the container, remove its stopper and flame the top. Insert the operating end of the pipette, mix the culture by part filling and emptying the pipette several times and then finally remove the sample. Reflame the top of the container and replace the stopper.

Removal of Organisms Growing in or on Solid Medium

(a) From Slopes: Slope cultures are made either in test tubes or bottles, hence similar points of aseptic technique are observed here as in the cases of liquid cultures. The tube is held in the same way in the left hand so that the surface of the solid culture is uppermost. The loop is flamed, the stopper is removed and the cool sterile loop is allowed to touch the surface of the slope. A small quantity of the growth is picked up on the loop, which is then removed from the tube. The mouth of the lid is reflamed and the lid replaced.

Any condensation water at the base of the slope will contain a large number of organisms and so alternatively some of this can be removed in the same way as for liquid culture, using a loop or pipette.

In neither case is the loop put down until subsequent inoculation operations are finished and it has been sterilised.

(b) From Plates:

Using a loop: When sub-culturing from plates a well isolated colony is chosen. The loop is sterilised and cooled. The plate is placed on the bench and twisted round so that the selected colony is in position convenient for the right hand. The right hand side of the lid is cautiously lifted with the left hand and the sterile loop used to pick off a portion of the isolated colony. It is easier to carry out this delicate operation if the side of the hand is rested on the bench in much the same way as in writing. Close the lid of the plate as soon as possible. If a colony is very small the whole of the colony may be removed.

Using a straight wire: Alternatively, if colonies are very close together and it would be difficult to remove a part, or the whole, of a colony without touching neighbouring colonies, or if a very small inoculum is required, then a straight wire is used in the same way as using a loop. Sub-culturing from colonies growing *in* solid medium is not an easy matter. Fortunately the occasion does not arise very often. A sterile scalpel loop or wire is used to cut out the agar containing the colony and to transfer it to the next vessel.

(c) **From Deep culture:** Anaerobic organisms are sometimes isolated from mixed cultures of aerobes and anaerobes by growing them in deep culture (See page 59). The anaerobic organisms will be found in the depths of the medium at the bottom of the test tube. To sub-culture them the following procedure is adopted.

Use a glass file to make a deep scratch around the tube about ½-¾ inch from its base. The outside of the tube is rubbed with alcohol. At the same time a pair of forceps is sterilised. The base of the tube is then passed through the flame several times and given a sharp tap with the forceps. The glass should crack all the way round on the score mark. Do not overheat the tube in the flame. With the forceps, and holding the whole lot over a jar of disinfectant, slip off the base of the test tube and discard it directly into the disinfectant. The solid medium, now projecting from the cut end of the tube and containing colonies of anaerobic organisms should be cut off with a sterile scalpel and dropped into a sterile container, such as a petri dish. The unwanted tube is discarded into the disinfectant. Further 'cutting' operations for inocula can be carried out safely in this way.

Inoculation of a Liquid Medium

(a) **Using a loop or straight wire:** The loop or straight wire containing the organisms is retained in the right hand. The tube, bottle, flask or plate from which it is derived is placed out of the way and the medium which is to be inoculated is moved into position.

Take the tube of sterile medium in the left hand and hold it in the usual way, but so that the tube is slightly tilted away from the vertical. Remove the stopper, flame the neck of the tube, introduce the wire loop and rub it against the glass of the tube in a position which is at the surface of the medium *at the moment* but will be below it when the tube is quite vertical. Surface tension forces will help to pull the organisms off the loop. The loop is removed carefully, the neck of the tube flamed, the stopper replaced and the loop sterilised in the flame.

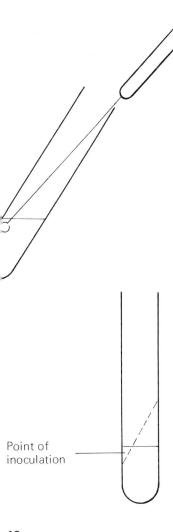

Point of inoculation

18
...culation of a liquid medium

The method is the same no matter where the organisms are derived. If they are derived from solid culture often a 'cloud' will appear in the medium at the point of inoculation. (See Fig. 18).

(b) **Using a pipette:** Use the mouth piece to control the flow and drop the inoculation into the liquid medium in such a way that the pipette itself *never* touches the liquid medium. Remove the stopper of the container and flame the top and then replace the tube to be inoculated in the rack. Introduce the pipette holding it vertically taking care not to touch the insides of the tube. A few millimetres above the medium expel the inoculum. The last drop may be removed by gently touching the inside of the tube. Remove the pipette, flame the top and replace the stopper.

If standardised pasteur pipettes are being used and a standard inoculum is desired care must be taken that the pipette is held absolutely vertical while the drop is being expelled.

The used pipette is discarded and immersed in a disinfectant bath or into a boiling water bath.

Make quite certain that newly inoculated cultures are well labelled.

It may be useful to note here that when tubes are being manipulated it is often possible to hold both the culture tube and the medium to be inoculated in the left hand at the same time. The two tubes are held so that one rests against the index finger and one against the middle finger and held jointly with the thumb. The two fingers are held slightly splayed out so that when the top of one is sterilised by flaming the stopper of the other does not catch fire. If this technique is adopted it will be found that it does speed up inoculation. It is not very easy to do with bottles because of the difficulty in top removal.

Inoculation of Solid Media

(a) **Inoculation of slopes:** The bottle or tube containing the nutrient slope is held in the left hand as described before, the loop containing the inoculum in the right. The right hand is also used to remove the stopper, and the mouth of the container is flamed. In the case of bacteria and yeasts: the loop is introduced to the base of the slope and a wavy line made on the slope with the loop gradually rising to the top. The loop is withdrawn the top flamed, the stopper replaced, and the loop sterilised. In the case of mould: a spot inoculum is made in a central position on the slope. (See Fig. 19).

(b) **Inoculation of stabs:** The organisms are on the tip of a wire, (and along some of its length if taken from a liquid culture). The

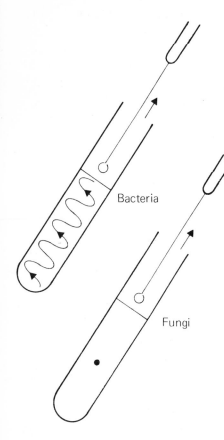

Fig. 19
Inoculation of slopes

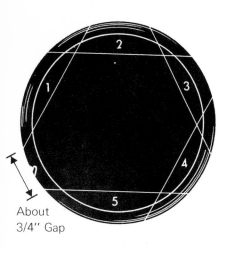

About
3/4" Gap

Fig. 20
Marking of plates for
inoculation

stopper of the tube of culture medium is removed, the top flamed and the straight wire plunged into the medium in a straight central vertical line, and then withdrawn. The neck is flamed, the stopper replaced and the wire sterilised.

(c) **Inoculation of plates:** In order to obtain isolated colonies of bacteria on a plate use one of the following methods.

Using a loop: Method 1. On the base of the plate draw segments with a grease pencil round the edge of the plate, slightly overlapping each with the one before, leaving a distinct gap between the first and the last. Number each in order for these act as the inoculation guide lines. (See Fig. 20).

Lift the right hand side of the lid with the left hand, and holding the loop so that its ring is almost parallel with the surface of the agar rub it up and down and completely cover the surface of segment one. Remember that it is easier if you rest the side of your hand on the bench. Remove the loop, flame it and at the same time replace the lid of the plate. Allow the loop to cool before you use it again. Twist the plate round anticlockwise and bring segment two into position. Then use the sterile loop in the same way as for segment one taking care that the organisms deposited in the overlapping portion of the two segments are spread over the whole of segment two. Sterilise the loop and allow it to cool, twist the plate round and carry on with the other segments until the last is completed. Do *not* run the last into the first. When the last segment has been spread extend into the unspread central area. Replace the lid, and flame the loop before you put it down.

Gain proficiency using transparent media through which the guide lines can be seen—before opaque media are used. (See Fig. 21). The type of growth that should be obtained is shown in Fig. 22. Well isolated colonies should be obtained in the central portion and probably also segments 5 and 4.

Using a loop: Method 2. On the base of the plate draw the guide lines as indicated in Fig. 23. Turn the plate the right way up, open it with the left hand as before and inoculate the narrow portion using a loop. This is the primary inoculum. Make certain the whole of the segment is inoculated. Remove the loop, close the lid and sterilise the loop. With the cool loop draw organisms along the three lines to the far side of the plate moving from the primary inoculum in each case. (Secondary inocula.) Sterilise the loop, close the lid and allow the loop to cool. Reopen the plate. In the large portion of the plate *and* leaving the middle portion entirely uninoculated draw a series of parallel lines moving in alternate directions. Then another series at right angles, and then two sets at the diagonal. Replace the lid and flame the loop. (See Fig. 24).

Using a pipette: Pipettes are used to transfer liquid cultures. If a liquid culture is used as an inoculum for solid medium these points

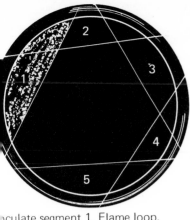

culate segment 1. Flame loop.

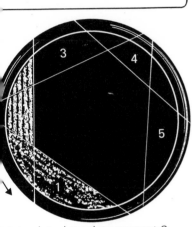

otate plate. Inoculate segment 2.
ame loop. Rotate plate.
oculate segment 3, etc.

ontinue rotating the plate anti-
ockwise to the final position
ig. 21
ethod for spreading a plate to
otain isolated colonies (method 1)

have to be observed:

Well dried plates must be used (2 hour dried) so that the liquid can be absorbed into the medium, depositing the organisms in many places over the surface of the plate. If a well dried medium is not used the surface of the medium will not dry after inoculation and confluent growth will result instead of isolated colonies. The pipette should be held quite vertical when the inoculum is expelled, and a mouth piece is used with it. The size of the inoculum should not be in excess of 0·1 cm

Used pipettes are discarded into disinfectant or direct into a water bath for boiling.

Using a spreader: This can be used when a plate has been inoculated using a pipette. The spreader is sterile and wrapped in Kraft paper, and is only unwrapped at the very last minute. The plate, inoculated from a pipette, is spread using the short end of the spreader which is pushed back and forth in straight lines from one side to the other, the lid being held in the left hand and the plate rotated with spare fingers until the whole plate is spread. The spreader is discarded into a water bath for boiling or into disinfectant. Before such a plate is incubated it is left on the bench 15 minutes or so while the moisture of the inoculum soaks into the medium. Then it can be inverted and incubated.

Using a large loop: 'Spot inoculation' Spot inocula are superimposed on plates already spread. The loop is charged, the handle held in a vertical position, so that the actual loop is horizontal, and touched down on the medium so that the inoculum discharges. The loop is then removed and sterilised in the flame.

B. Cultural Methods

All cultures should be labelled clearly with the following information: Name of pupil; name of organism or material cultured; medium; time and temperature of incubation; date.

When one or more species are growing together in or on one medium the culture is said to be *mixed*. This may be obtained for example, when dust is sprinkled on a plate of nutrient medium which is then incubated, or if a plate is exposed to the air.

The term *pure culture* refers to the growth of a single species in a culture medium. Cultural methods rely on obtaining pure cultures so that the characteristics of a species can be observed without confusion.

Pure cultures are obtained by taking two steps:

    obtaining isolated colonies,

    separating these isolated colonies and thus obtaining pure cultures.

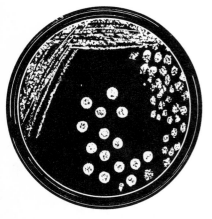

Colonies, small and crowded— not typical.

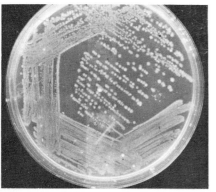

Note large isolated colonies in the central portion.

Fig. 22
Showing the type of growth after inoculation by method 1

1. Methods for Obtaining Isolated Colonies

Using a loop: These methods have already been dealt with earlier in this chapter, and there is no need to reiterate them here.

Pour plate method: Essentially this method comprises the serial dilution of a liquid culture in a sterile melted, cooled, normally solid, medium, the whole poured into a petri dish, allowed to set and incubated. Colonies grow embedded in and on the surface of the medium, and in those plates which contain sufficiently dilute inocula there will be isolated colonies.

The culture: Either a previously inoculated and incubated liquid culture, or a test sample (soil suspension, milk, etc.) is used. The quantity of growth will determine how many dilutions are set up . . . a lot of growth demands a series leading to high dilution, a small amount of growth a series leading to a low top dilution. The range of dilutions may have to be determined by trial and error.

Making the dilutions: The diluent used is sterile, melted cooled (to about 45-48°C) nutrient agar or other nutrient medium. Three or four tubes will be required. They must be inoculated and kept melted until you are ready to pour them as plates. Transfer three large loopfuls of the neat culture into the first tube of sterile melted medium. Sterilise the loop. Rapidly mix this tube by rotating it between the hands, but take care not to froth it. Withdraw three loopfuls from it and transfer them into tube two. Sterilise the loop. Mix the tube as before. Transfer three loopfuls from this tube into tube three. Sterilise the loop and mix the tube. Keep all inoculated tubes in the melted condition.

Pour each tube of melted agar into a separate sterile petri dish, labelled with the appropriate dilution. Incubate either at room temperature or in the incubator at 37°C.

Results: It should be found that the plates inoculated with the higher dilutions have isolated colonies, some of which will be on the surface, most of which will be embedded in the agar.

Spreader method: In principle this method is very similar to the pour plate method. Dilutions are made as before, but using ¼ strength Ringers solution as the diluent. The agar plates are poured and well dried for two hours before inoculation. A small quantity of the dilution (0·1 cm³ at most) is added to the surface of the plate from a fine pasteur pipette. The drop is spread using a spreader (see page 53) and before incubation the moisture of the inoculum is allowed to penetrate the plate. After incubation the higher dilutions will show isolated colonies on the surface of the plate.

Shake culture for the isolation of anaerobic organisms: An inoculum of, for example a suspension of soil, is added to a bottle or tube of cooled melted nutrient agar. The whole is well mixed and allowed to gel in its container. Anaerobic organisms will be

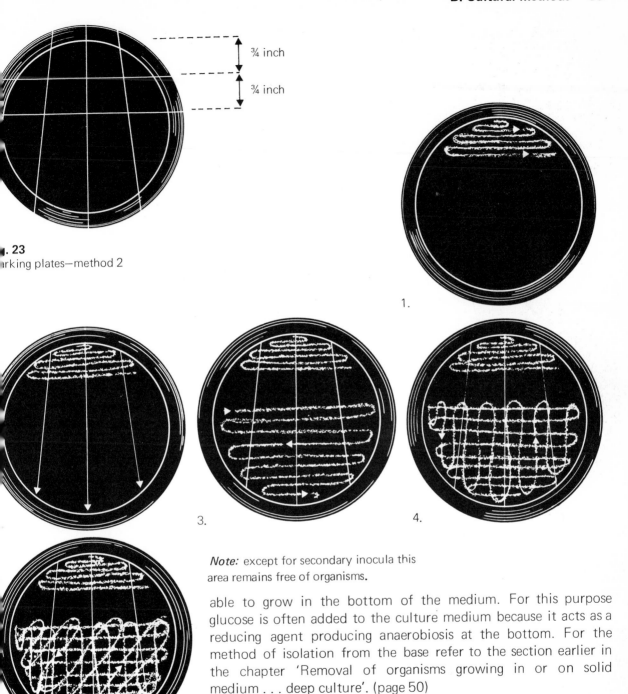

¾ inch

¾ inch

Fig. 23
Marking plates—method 2

1.

Note: except for secondary inocula this area remains free of organisms.

3.

4.

Fig. 24
Steps in inoculating a plate by method 2

able to grow in the bottom of the medium. For this purpose glucose is often added to the culture medium because it acts as a reducing agent producing anaerobiosis at the bottom. For the method of isolation from the base refer to the section earlier in the chapter 'Removal of organisms growing in or on solid medium . . . deep culture'. (page 50)

Use of selective media: Often selective media are used to select particular species from an inoculum in which unwanted species are predominant. Reference will be made to these methods in the next few chapters in relation to the examination of water, milk, soil and so on. pH is also a selective factor since many bacteria favour a pH of around 7·4, while a pH of around 5·4 favours the growth of moulds and fungi.

2. Pure Culture

Isolated colonies growing on plates are picked off carefully using a sterile loop and inoculated into sterile nutrient broth, or other suitable liquid medium. This is then incubated at a suitable temperature and allows for the multiplication of the species. Its purity is then checked by inoculating a plate to get isolated colonies. If the culture is seen to be pure, then a representative colony is used to make a stock culture.

3. Stock Culture Methods

These are methods by which stocks of different types of organisms are maintained in the laboratory. Since micro-organisms multiply so quickly the number of generations in a given time is large. The way in which stock cultures are set up is designed to minimise the chance of selection of mutants and so maintain the species as true to form as possible. There are several ways of keeping stock cultures, one of the easiest being on slopes of nutrient medium contained in screw cap bottles.

Slopes: Slopes are inoculated in pairs according to the scheme shown in **Fig.** 25 and are sub-cultured ('subbed'—that is transferred to fresh medium) regularly every four to six months, or sooner if necessary. The stock culture should be stored in a cool dark cupboard or at 4°C in a refrigerator. After inoculation the slopes are incubated for a short time, 8-12 hours, a time long enough to produce just visible growth on the surface of the medium. The bottles should be well labelled in a permanent way using self adhesive labels, labelling one *stock* and the other *routine*. The *routine* bottle is used to set up cultures for experiments, the stock bottle is used *only* to inoculate the next pair of bottles when the strain is subbed.

Freeze dried cultures, reconstitution: Cultures obtained from national collections will probably be freeze dried and contained in an ampoule.

How to open an ampoule and regenerate a freeze dried culture:
Assemble the following:

   The ampoule of culture

   A glass file or diamond pencil.

   Glass rod

   Forceps

   Sterile pasteur pipette and mouth piece carefully unwrapped and connected to the mouthpiece.

Bunsen burner

Base of a petri dish

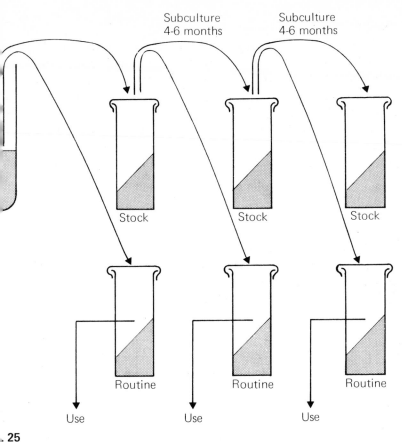

Subculture
4-6 months

Subculture
4-6 months

Stock   Stock   Stock

Routine   Routine   Routine

Use   Use   Use

25
owing the scheme for
ing up slope stock cultures

Test tube rack containing 2 test tubes (sterile) each containing about 5 cm³ nutrient liquid medium, (selected according to species of organism).

Do the following:

1. With the diamond pencil score the glass of the ampoule half way along the cotton wool plug.

2. Heat up the end of the glass rod to red heat and apply it to the score mark on the ampoule. The glass should crack on the line.

3. Holding the base of the ampoule, with sterile forceps (flamed) remove the top of the ampoule and the cotton wool plug. Place these two in the prepared petri dish base. Sterilise the forceps and put them down.

4. Carefully pick up the pasteur pipette and mouthpiece. Remove the stopper from one test tube of nutrient broth and withdraw 1·5 cm³ or so. Transfer this with great care into the open ampoule (still held in your left hand). Mix well by expelling

and withdrawing several times and then finally withdraw it al
from the ampoule and transfer it aseptically to the 2nd tube o
nutrient medium.

5. Dispose of the pasteur pipette into a boiling water bath, anc
treat it appropriately according to whether it is a culture of spore
bearing organisms or not. Put the empty ampoule into the half petr
dish and immerse the whole in strong disinfectant for several days
(See Chapter 2.)

6. Incubate the broth, and proceed from the broth culture
Once checked for purity slope cultures for stock can be set up.

Cultures on blotting paper discs: It is possible to keep cultures
on a short term basis on blotting paper discs or squares.

Cut up a piece of fairly thick blotting paper into pieces abou
0·8 cm square. Put them into a stoppered bottle or test tube anc
sterilise them in the hot air oven. Impregnate these squares with
suspensions of dense cultures either by dipping them or by addin(
a drop of culture from a pasteur pipette. Dry the squares b)
putting them in a sterile petri dish and placing them in the
incubator for a short time. Store them dry in a screw cap bottle ir
a cool place. Label each bottle clearly with the organism and th(
date. This can be a very quick way of setting up a lot of culture:
of an organism.

Where to obtain cultures:

1. National collections. See appendix.

2. Local hospitals. Contact the chief technician of the
Pathology Laboratory, who may be able, and willing, to help
you.

Always state why you want the organisms.

3. Isolate your own.

A   B   C   D   E

Fig. 26
Streak plate

4. Streak Plates and Lawns

Streak plates: Several different species of organism can be
grown on the same plate if they are inoculated in parallel lines or
the surface of the medium. It is a good idea to first of all mark anc
label the base of the plate with a grease pencil to indicate the
positions of the cultures. (See Fig. 26)

Confluent growth or 'lawns': In some experiments it is
necessary to have growth of the bacteria covering the entire
surface of the plate, this is known as a lawn.

A fairly dry plate should be used, in conjunction with a large
inoculum dropped onto the plate from a pipette. This is
distributed over the whole surface with a spreader. Excess
moisture should be absorbed by the plate leaving it slightly wet

looking. After incubation there should be growth over the whole plate with no separate colonies.

5. Anaerobic Methods

Not all micro-organisms require oxygen, indeed some can only grow if the quantity of oxygen in the surrounding atmosphere is limited (micraerophilic), some can only grow in the entire absence of oxygen, the anaerobes. These organisms are quite common and may be isolated readily from soil. Eliminating oxygen from the atmospheric environment surrounding a culture is not an elaborate procedure.

Shake tubes or deep agar: A long column of solid nutrient agar (20 cm^3 in a test tube) will supply anaerobic conditions at its base where, once inoculated, the anaerobic organisms will grow. The medium is inoculated either using a straight wire, or by melting and cooling the medium and introducing the organisms using a pasteur pipette.

Glucose broth in a long test tube. If it is heated in a boiling water bath for ten minutes or steamed for thirty minutes it will provide an anaerobic medium. The medium is cooled and sterile melted vaseline poured onto the top to seal it. Inoculation is through the melted vaseline using a pasteur pipette.

Robertsons cooked meat provides an excellent medium for the growth of anaerobes since the constituents of the meat are reducing agents and maintain the anaerobic conditions (see page 29).

Sterile strips of iron or nails added to nutrient liquid media previously heated in a water bath for ten minutes or steamed for 30 minutes will maintain anaerobiosis. The iron strips can be sterilised in a bottle in the hot air oven and in which they will subsequently be kept.

Use of film impervious to oxygen: Anaerobiosis is achieved by addition of an aerobic micrococcus to the nutrient medium itself. *(Micrococcus albus)* After anaerobiosis is achieved the micrococcus dies and leaves the anaerobe in a pure state. To prevent the diffusion of oxygen into the medium impervious film is used.

Micrococcus albus is grown on whey agar slope containing 2% calcium carbonate, and incubated at 20°C. The growth is scraped off the surface with a loop.

Nutrient medium for the anaerobe: Any suitable medium can be used, preferably one not rich in carbohydrate (to reduce gas evolution) with a pH of between 5 and 7. Autoclave to sterilise it.

Indicator: To indicate a state of anaerobiosis resazurin has been found to be suitable. A stock solution of 0·5% is required.

Film: This must be impervious to oxygen. Cellophane, poly¹
thene and so on are not suitable. Suitable vinylidene chloric
polymer can be obtained from the Dow Chemical Co, Midlan¹
Michigan : Saran A, 100 gauge, 0·02 mm thick, 'pre-shrunk'; ¹
from Messrs Plasfoils, Portland House, Stag Place, London S.W.

Sterilisation of film: Immerse the film for about 1 minute in
solution composed of:

60 parts alcohol (96%)
30 parts water
10 parts concentrated hydrochloric acid.

Rinse the film by agitating it gently in four changes of steri¹
water, remove slowly from the last change, drain it and immed¹
ately place on the plate.

Anaerobe: If this is being isolated from a food such as a mea
sample pasteurise it first by heating it to 90°C for 10 minutes t
destroy vegetative cells. Make a series of dilutions of it and us
these to inoculate the plates.

Procedure: Grow a culture of *Micrococcus albus*. Make th
nutrient medium for the anaerobe and autoclave it. Cool it t
about 45°C and add enough of the micrococcus to achieve a sligh
turbidity. Add indicator to about 0·0005% concentratio
(approximately 1 drop of 0·5% solution in 40 cm³ medium
Inoculate the still liquid agar medium with 1 cm³ of dilution ¹
the anaerobe. Pour the plate and allow it to set and to dry wel
When dry press the sterile film onto the surface. Incubat
aerobically. Colonies of the anaerobe will develop under the film

Growth with a strongly aerobic organism: Anaerobic organism
can be grown on ordinary nutrient media in the following wa¹
One half of the well dried plate is inoculated with a strongl
aerobic organism such as *Micrococcus albus.* The other half
inoculated with an anaerobic organism. The plate is then seale
round the rim with vaseline. The aerobic organism grows an
thrives while there is oxygen enough for it inside the plate. Whe
the oxygen is exhausted it ceases to grow but the anaerobi
organisms can then grow easily.

Anaerobic jar: This is a jar of stainless steel or thick glass int
which cultures are placed. The jar is sealed and then evacuated ¹
its air which is subsequently replaced with another gas, usuall
hydrogen. A tube of indicator is used to ensure that anaerobi
conditions are achieved and maintained.

Precautions:
1. The jar must not be overfilled since this would impede th
evacuation of the jar.
2. The lids of bottles and plates must be loose to allow th
evacuation of air from their insides. Care must be taken that th

lids of petri dishes are loose otherwise condensation water can form a seal at the edge of the plate and allow local aerobic conditions.

Method of use:
1. Fill jar.
2. Close lid firmly.
3. Close inlet valve.
4. Attach jar to evacuating pump and evacuate until the internal pressure is about 60 mm Hg.
5. Close valve between pump and jar.
6. Using a reducing valve fill a balloon or rubber bladder, whose total capacity is greater than the capacity of the jar, with hydrogen from a cylinder.
7. Connect the bladder of hydrogen to the inlet valve of the jar. Open the valve and allow the jar to fill with hydrogen. After this some hydrogen should be left in the bladder.
8. *Either:* Place the jar and bladder in a large wooden box. This is a precaution against explosion. Attach the terminals of the jar to the mains. Close the lid of the box, and then turn on the current. Any residual oxygen in the jar will be converted to water which will condense and so more hydrogen will be drawn in from the bladder. Turn off after about twenty minutes. Open box. Disconnect from the power supply.
Or: Use a cold catalyst. This catalyses the reaction $2H_2 + O_2 \rightarrow 2H_2O$. The water which forms condenses and more hydrogen is drawn in from the bladder until catalysis is complete. It is a much safer method.
9. Close the valve between the bladder and the jar, and disconnect the bladder.
10. Place the jar in the incubator. If anaerobic conditions have been attained the tube of indicator will remain colourless.

Indicator: Mix in equal volumes the solutions *A*, *B* and *C*.
A 6 cm³ 0.1N NaOH to 100 cm³
B 3 cm³ 0.5% aqueous methylene blue to 100 cm³
C 6 gm glucose to 100 water. Add a small crystal of thymol.
Boil the solution to render it colourless and place approximately 10 cm³ in a test tube inside the jar. (See Fig. 27).

Culture in an atmosphere of increased carbon dioxide concentration: This can be done in a very simple way. The cultures are arranged on the base plate of a bell jar together with a lighted candle. The rim of the bell jar is vaselined and then placed over the cultures and candle and a firm seal secured.

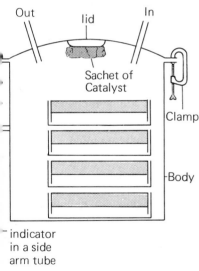

Out   lid   In

Sachet of
Catalyst

Clamp

Body

indicator
in a side
arm tube

g. 27 Anaerobic jar

te: Many jars are available which
e cold catalysts—these are *much*
fer.

C. Incubation
All agar plate cultures of **bacteria and yeasts** are incubated *agar*

side up—this obviates condensation water dropping onto the plate and consequent confluent growth. All agar plate cultures of *moulds* are incubated *agar side down* preventing spores falling into the lid and later being disseminated about the laboratory.

No incubator: If lack of money or other reasons prevent purchase of an incubator micro-organisms can be grown fairly satisfactorily at room temperature. It is preferable to place the cultures in a position in the room where the temperature fluctuates least, where they will not be disturbed by other pupils using the laboratory, or by accident, and where the risk of contamination of the cultures is at a minimum. Use of a warm cupboard would be all right provided that certain precautions are taken. There is always a risk that whatever you use as an incubator will get contaminated by cultures being dropped or spilled so it is better to use a cupboard whose inside surfaces are smooth and easy to disinfect. For this reason a metal cupboard is preferable to a wooden one.

Note: If you do adopt this procedure which is far from ideal do not use the cupboard for any other purpose either at the same time or at any other time. Thus a fairly small one—2 feet square with two or three shelves should suffice.

Cultures should always be kept out of the way of the uninitiated.

Home made incubator:

It is possible to make your own incubator at very low cost and very easily.

Purchased incubator:

There are many makes of very reliable incubators on the market. An incubator should be well lagged and preferably, though not essentially, be provided with two doors, an outer lagged door and an inner glass door. The thermostat should be sufficiently sensitive to maintain the pre-set temperature with little fluctuation. For most school purposes the finest temperature control is not essential, but it should be able to maintain 37°C fairly easily.

D. Preservation of Plate Cultures

1. Use of formaldehyde as a fixative

A drop of formalin in the lid of a plate will kill the organisms growing on that plate. The plate should be exposed for at least ten minutes. This process will not alter the appearance of the plate.

This procedure could be used before plates are put out before a class for demonstration purposes.

2. Refrigeration

Plates can be preserved in a refrigerator at 4°C. Provided that they are not allowed to freeze or dehydrate the plates can be preserved for several weeks and often longer. This is invaluable. If a class set up an experiment one day growth will occur overnight and should be observed the following day. Often this is not possible. Placing the plates at a certain stage of development in the refrigerator will mean that they can be observed and the right results recorded when the class next meets.

3. Sealing plates with vaseline

This will prevent the dehydration of the medium. Once the atmosphere inside the plate has been exhausted or is noxious growth will stop and the plate can be preserved without refrigeration.

4. Use of clear resin

A layer of clear embedding resin such as 'Bedacryl' (I.C.I.) can be poured over the surface of a plate culture in a *glass* plate (plastic dissolves) and is a useful and safe means for keeping cultures for display.

E. Counting Methods

The number of organisms in a suspension can be assessed in two ways.

One method is to observe a thin film under the microscope and to count all cells in a given area of known depth. A calculation is done which gives the **total count** per millilitre of the original suspension. This method does not differentiate between living and dead bacteria.

The other method is to make serial dilutions of the test suspension and to inoculate a small volume of the dilutions onto a solid medium and to incubate. Colonial growth occurs. Every colony appearing on the solid medium is counted and is taken to represent one cell from the original suspension. So by calculation the number of living cells per unit volume can be assessed. This method is a **viable count** since only living cells give rise to colonies.

Each of these methods is open to error since small samples are taken to be representative of large volumes. In the total count cells may clump together, and detritus may be mistaken for cells; in the viable count single colonies may arise from 1, 2 or from a clump of several cells. Not all the cells present will grow—only those capable of growth in the prevailing conditions.

Total Count

A slide with a Thoma type grid engraved on it is used. Special slides designed for counting bacteria are available although haemocytometers with the improved Neubauer grid can be used. The slide is of thin glass and in the central third a circular ditch is cut. The platform contained within this is 0·02 mm lower than the rest of the slide. The counting grid is engraved in its centre. The grid is 1 mm square and is divided into 400 very small squares each having an area of 0·0025 mm². A small drop of the test suspension is placed over the grid with a loop. The optically plane cover slip is placed over the central platform and is carefully pressed down until Newton's rings are seen. Great care must be taken that the suspension is not allowed to over run into the ditch because if this happens the depth between the coverslip and grid will not be accurate. When the coverslip is correctly in position the depth between the grid and the lower surface of the coverslip will be 0·02 mm. Hence the volume over each small square will be:

$$= 0{·}02 \times 0{·}0025 \text{ mm}^3$$
$$= 0{·}000\ 000\ 05 \text{ cm}^3$$

(See Fig. 28)

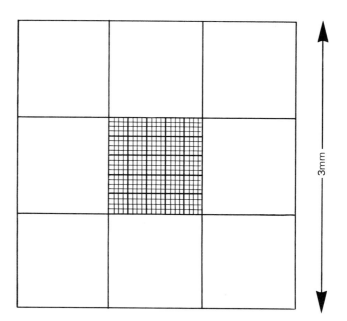

Fig. 28
Thoma type grid

Equipment 1 young (12-18 hour) broth of
        B. cereus (or similar)
        0·1% methylene blue . . . filtered
        40% formaldehyde solution plus dropper
        1 bacterial counting chamber plus its cover slip
        (or haemocytometer.)

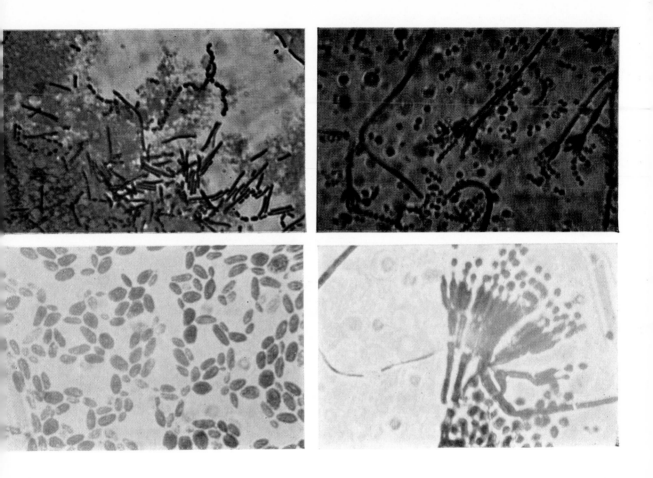

Top Lactobacilli—large square-ended rods, and streptococci in yoghurt. (Stained in neutral red: mag. ×3000, oil immersion.) (c.NCFT)

Bottom Cells of the yeast *Saccharomyces cerevisiae* showing budding. (Mag. ×800. Stained with methylene blue) (c.NCFT)

Top Penicillium species (Mag, ×200; stained lactophenol blue) (c.NCTF)

Bottom A penicillius of Penicillium species. (Mag. ×800; stained lactophenol blue) (c.NCFT)

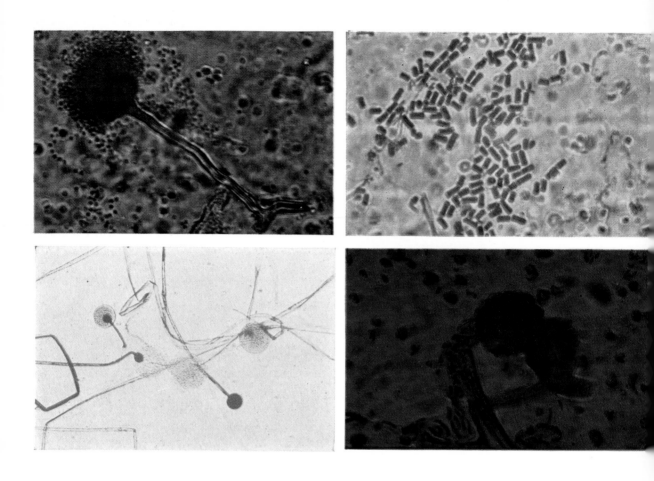

Top Aspergillus species—showing conidial head and foot cell. (Mag. × 200; stained lactophenol blue) (c.NCFT)

Bottom Sporangia at all stages of formation, in a species of the Mucorales. (Mag. × 200, stained lactophenol blue) (c.NCFT)

Top Arthrospores of the mould Geotrichum in yoghurt. (Mag. × 200, in lactophenol blue) (c.NCFT)

Bottom Zygospore formation in a species of the Mucorales (Mag. × 800; in lactophenol blue.) (c.NCFT)

1 test tube

1 tube isotonic saline

Method: Fix the bacterial suspension by adding 2 drops of 40% formaldehyde solution. Mix thoroughly by 'flicking' the base of the firmly held tube. Stain the bacteria by adding a drop of the methylene blue, this makes them easier to see.

Rinse wash and dry the counting chamber and its cover slip.

Place a small loopful of solution in the centre of the chamber platform and apply the cover slip. The size of the loopful must be just enough to fill the platform but no more. Press down the cover slip until Newton's rings are seen uniformly distributed in the area of contact. Examine using the 'high dry' objective with the iris diaphragm closed.

Method for counting: The total field comprises 25 large squares each containing 16 small squares. Take a piece of graph paper and draw a large grating on it, and use it systemmatically to record counts per square. (See Fig. 28) Record the number of cells in each of the small squares counting enough squares, about 100, to give a count of cells between 300 and 1000.

Adopt the convention that any cell which overlies the top or right hand lines of the square is *included* in that squares total, and that any cell which overlies the bottom or left lines of that square is *ignored.* (See Fig. 29)

Calculation: Calculate the average number of cells per small square (n)

$$n = \frac{\text{total number of cells}}{\text{number of very small squares counted}}$$

Volume of each small square = $0.000\ 000\ 05\ cm^3$

Therefore $0.000\ 000\ 05\ cm^3$ contains n cells

And $1.0\ cm^3$ of the original suspension contains $\frac{n}{5} \times 10^8$ cells

$$= n \times 20\ 000\ 000$$

If the original suspension has been diluted multiply also by the dilution factor to obtain the final answer.

Repeat the count on three separate samples and average the answers.

Viable Counts

These methods count only living bacteria

Pour plate method

Equipment: 9 sterile test tubes

1 sterile graduated $10\ cm^3$ pipette

1 young broth *Staph. aureus*, or fluid sample such as water, milk and so on.

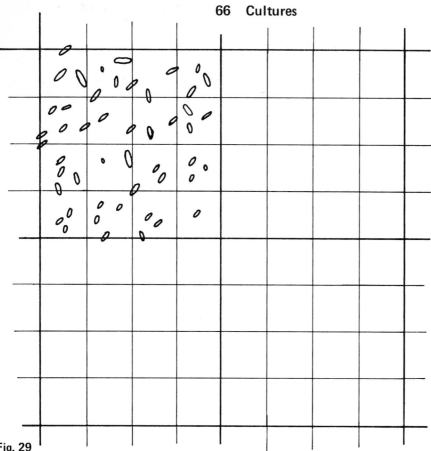

Fig. 29
To illustrate the convention
in counting cells

3 4 1 2
4 3 3 4
4 2 2 3
4 4 2 1
Total: 46

Convention. Include cells touching
or overlying top and right sides of
any square. Ignore any cell touching
or overlying the bottom or left hand
lines of any square.

9 1 cm³ sterile pipettes
12 petri dishes
12 bottles sterile melted cooled N.A.
¼ strength Ringers solution (sterile)

Method: Prepare serial dilutions (See page 69): Pipette 9·0 cm³ amounts ¼ strength Ringers into each of the 9 sterile tubes using the 10 cm³ pipette.

Mix the bacterial suspension well.

With a sterile 1·0 cm³ pipette transfer 1·0 cm³ of suspension into the first test tube of diluent. With a *fresh* 1·0 cm³ pipette mix the 1st dilution by filling and emptying the pipette several times, and then transfer 1·0 cm³ of this into the second tube of diluent. Make the remaining dilutions in the same way, using a fresh pipette for each.

Starting with the **greatest** dilution, pipette $1 \cdot 0$ cm^3 amounts of each dilution into each of three sterile petri dishes. Then pour into each dish about 10 cm^3 (not less) of clear N.A., melted and cooled to 45-50°C. Mix at once by rapidly moving the plate, while flat on the bench, in a combination of sideways and circular movements in alternately clockwise and anti-clockwise directions. Take care that the N.A. is not spilt on the bench.

Incubate at 37°C for 24-48 hours. Make certain that each plate is well labelled with the dilution of the inoculum e.g. 1/1000 or 1/10,000 and also with your initials.

Result: Count the colonies in the three plates which give between 50 and 500 colonies per plate.

Calculation: Suppose the following dilutions have been made: neat, 1/10, $1/10^2$, $1/10^3$, $1/10^4$, $1/10^5$, $1/10^6$, $1/10^7$, $1/10^8$, $1/10^9$, and that $1 \cdot 0$ cm^3 from the last four dilutions have been plated in triplicate a possible set of results might be:

Colonies on each plate:		$1/10^6$	$1/10^7$	$1/10^8$	$1/10^9$
	1.	421	42	4	1
	2.	398	38	5	0
	3.	402	40	3	2
Total:		1221	120	12	3
Average per plate		407	40	4	less than one

Since the count for the $1/10^6$ plates falls between 50 and 500, the calculation is based on the result from this plate (but checked against the other dilutions) 1 cm^3 $1/10^6$ gives rise to 407 colonies that is equivalent to 407 organisms. Therefore 1 cm^3 of the original suspension contained $4 \cdot 07 \times 10^8$ organisms/cm^3.

The counts on the other plates should be approximately one tenth and one hundreth of the $1/10^6$ plate and are a check on accuracy. If there is a wide discrepancy in the results the technique used should be closely examined.

Note: Colonies will be growing on the surface of the agar and these appear large, they also grow *in* the agar and these will be much smaller.

Roll Tube Count

This method is fundamentally similar to the pour plate method for counting organisms except that the medium is allowed to set round the sides of a test tube instead of in a petri dish. It uses less agar and provided the technique is mastered is entirely satisfactory.

Equipment   9 sterile test tubes
1 sterile, graduated 10 cm³ pipette
Sterile isotonic saline
1 young broth *Staph. aureus*
9 sterile 1 cm³ pipettes
1 0·1 cm³ sterile pipette
15 6 in x 5/8 in test tube containing 2 cm³ sterile
NA. Melted and cooled to 45°C. *Note.* The agar shoul
be 2·0-2·5% that is, more concentrated than
usual. These will be the **Roll tubes.**

Method: Make serial decimal dilutions in the same way as
previously described. With a sterile pipette add 0·1 cm³ of the
highest dilution to one roll tube. Rapidly mix the culture into the
agar. Then hold the tube in a near horizontal position and rotate
the base of the tube in cold water, under a stream of water from
the tap, or in a groove in a block of ice. The agar must set in a thin
layer round the sides of the tube. Label the tube and proceed to
the others, inoculating and then rolling them. Incubate at 37°C
inverted for at least three days.

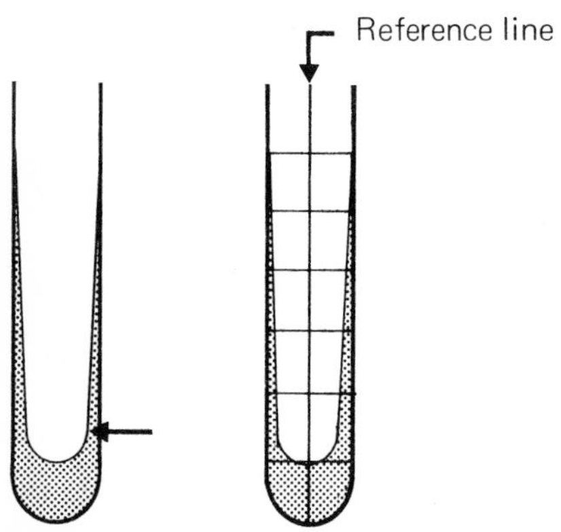

Results: Draw a vertical reference line down the tube and a
series of horizontal ones parallel down the tube. Record the
number of colonies in each segment of the tube and then total up
for each tube. Calculate the number of organisms in 1 cm³ of the
original suspension in the same way as for previous counts
remembering that the size of the inoculum was 0·1 cm³ (See
Fig. 30).

Footnote. This technique is only easy if roll tubes are used in
conjunction with an electric tube roller (*Astell.*)

Surface Viable Count

This method is used to count organisms which grow best on the
surface of the medium that is, ones which are strongly aerobic.

Equipment: 15 2 hour dried sterile N.A. plates
9 sterile test tubes
1 sterile graduated 10 cm³ pipette
1 young broth of *Staph. aureus*
Sterile isotonic saline or distilled water
9 1 cm³ sterile pipettes
1 loop

Method: Make serial decimal dilutions in the same way as for
the pour plate method. Then starting at the highest dilution
pipette 0·1 cm³ of each dilution onto the surface of each of three
plates and spread it widely over the agar with the loop. Leave the
plates standing on the bench for at least 15 minutes during which
time the water from the inoculum will be absorbed. Incubate at
37°C for 24 hours.

Reference line

Sectional view
of roll tube

Roll tube marked
ready for counting

Fig. 30
Roll tubes

Results: In this case there will only be growth on the surface of the agar. Count in the same way as for the pour plate method. Assuming the same sample is being used as in the pour plate method typical results might be as follows:

	$1/10^5$	$1/10^6$	$1/10^7$	$1/10^8$	$1/10^9$
	420	41	4	1	0
	398	38	5	1	0
	402	39	3	2	0
Average	407	39	4	1	0

The inoculum was 0.1 cm^3 (c.f. pour plate method)
0.1 cm^3 $1/10^5$ dilution gives rise to 407 colonies that is the equivalent to 407 organisms.
Hence 1.0 cm^3 of the original suspension contained

$$\frac{407 \times \text{dilution factor}}{\text{Size of inoculum}}$$

$= 407 \times 10^5 \times 10$ orgs per cm^3
$= 407\ 000\ 000$ organisms per cm^3

Enumeration by Probability Methods
See *'water'*, p. 127

Making Dilutions

The diluent used is sterile isotonic saline or N.B.
Decide on the series of dilutions wanted:
1/2, 1/4, 1/8, 1/16 and so on—'Doubling dilutions'
1/10, 1/100, 1/1000, 1/10 000 and so on—'Decimal dilutions'
Set up a series of sterile test tubes. The dilutions may be made accurately using a series of sterile pipettes or, roughly by tipping from one dilution into the next.

Rough decimal dilutions: 1/10 dilutions can be made by marking the empty tubes at the 9 cm^3 and 10 cm^3 levels. Then fill all the tubes up to the 9 cm^3 mark with sterile diluent. Add to the first tube of diluent culture to fill it up to the 10 cm^3 mark. This gives a 1/10 dilution. Mix this well and use some of it to make the second tube of diluent up to the 10 cm^3 mark so giving a 1/100 dilution. This procedure can be extended to give as many dilutions as desired.

Accurate decimal dilutions: Put 9 cm^3 exactly into a series of sterile dry empty test tubes using a graduated pipette and mouth piece. Make the decimal dilutions using a series of sterile 1 cm^3 pipettes as follows, using Fig. 31 for guidance.

Mix tube 1, the neat culture, then introduce 1 cm^3 pipette

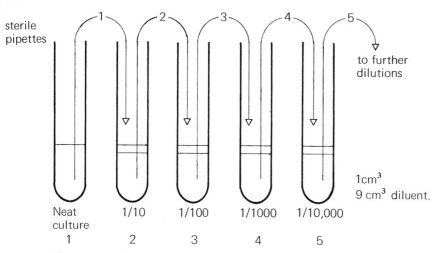

sterile pipettes

to further dilutions

1cm³
9 cm³ diluent.

Neat culture	1/10	1/100	1/1000	1/10,000
1	2	3	4	5

Note: In practice all tubes would have stoppers.

Fig. 31
Decimal dilutions

number 1 and *withdraw* 1 cm³. (Do *not* mix it with the pipette). Expel it into tube 2, holding the tip of the pipette slightly *above* the surface. Discard pipette number 1.

Sterile pipette 2: Mix tube 2 using this pipette. When well mixed withdraw 1 cm³ and expel this into tube 3 with the tip slightly above the surface. Discard pipette number 2.

Sterile pipette 3: Mix tube 3 using this pipette. When well mixed withdraw 1 cm³. Expel this into tube 4. Discard pipette number 3.

Sterile pipette number 4: Mix tube 4 with this pipette etc. Carry on in this way until the appropriate series has been made.

Doubling dilutions: 1/2, 1/4, 1/8, 1/16 and so on. Measure 5 cm³ diluent, or a known volume, into each of a series of sterile test tubes. To the first tube an equal volume of culture is added, either accurately with a pipette, or roughly. This tube, ½ dilution, is well mixed, and half of it is transferred to the second tube of diluent using a second sterile pipette, or roughly. This make a ¼ dilution of the culture. The process is repeated as often as is required. (See Fig. 32)

Quadruple dilutions: 1/4, 1/16, 1/64 and so on. Mark, and fill the tubes to the 6 cm³ level with diluent. To the first tube add 2 cm³ culture, giving a ¼ dilution. This is mixed and 2 cm³ of it transferred to the next tube giving a 1/16 dilution and so on. (See Fig. 33)

Other series: If a series such as 1/10, 1/20, 1/30, 1/40, 1/50, 1/60, 1/70 and so on is required the quantities of dilution fluid and culture to be mixed must be carefully calculated beforehand.

Sterile pipettes:

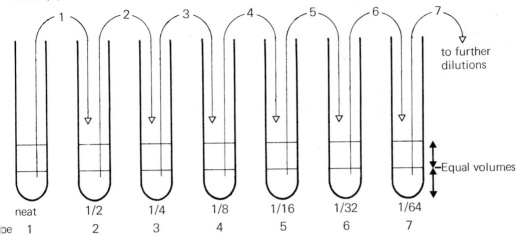

to further dilutions

Equal volumes

| neat | 1/2 | 1/4 | 1/8 | 1/16 | 1/32 | 1/64 |

| be 1 | 2 | 3 | 4 | 5 | 6 | 7 |

te: In practice all tubes would
e stoppers.

32
ubling dilutions

Sterile pipette
no:

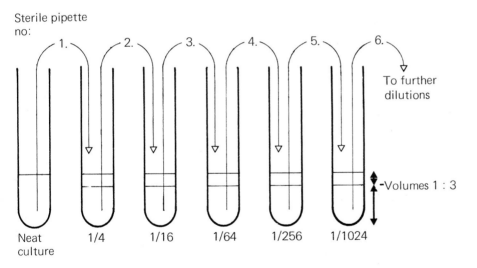

To further
dilutions

-Volumes 1 : 3

| Neat culture | 1/4 | 1/16 | 1/64 | 1/256 | 1/1024 |

te: In practice all tubes would have stoppers

. 33
adruple dilutions

In order to make a particular dilution remember the formula:
Volume of inoculum to add $= \dfrac{\text{Dilution wanted}}{\text{Dilution of stock solution}} \times$ Final containing volume

Counting bacterial numbers by opacity and turbidity

Browns opacity tubes

A set of glass tubes of varying opacity can be purchased. The test suspension of bacterial cells is compared with various of the tubes, until the one closest in opacity is found. Reference to a table—supplied with the tubes—gives the approximate number of cells/cm³. This can be quite useful as a guide to the number of dilutions to set up for a viable count.

Turbidimetric methods

When light is passed through a bacterial suspension the light is scattered. The amount of light scattered by the cells in suspension is proportional to the number of cells present. (This is true in dilute suspensions, but not true in dense suspensions). A *nephelometer* is used to measure the light scattered. This is simple to use, reliable and not too expensive. A *spectrophotometer* measures the amount of light transmitted—which again—in dilute suspensions—is proportional to the number of cells present. This is an expensive piece of equipment—and for this purpose not very reliable.

For each it is necessary to prepare a calibration curve relating the optical density (scale reading) to the number of organisms per cm³ of suspension. Once the calibration curve is prepared reference can be made to it to determine the number of cells per cm³ for any observed opacity.

Preparation of calibration curve: Take a young culture or organisms, for example, *E. coli* and prepare a series of dilutions of it in saline:

Tube number	Volume of culture cm³	Volume of saline cm³
1	1	9
2	2	8
3	3	7
4	4	6
5	5	5
6	6	4
7	7	3
8	8	2
9	9	1
10	10	0

Obtain (1) the *turbidity* of each tube on the nephelometer (or spectrophotometer) and record it, and (2) *either* make a series of decimal dilutions up to 1/10⁷ of the culture of organism and plate out 1 cm³ of each dilution in triplicate in nutrient agar using the pour plate method. Incubate all plates at 37°C for 24-48 hours.

From these obtain a *viable count* per cm³ of the original suspension of organisms. Use this to calculate the number of organisms per per cm³ of each of the ten tubes examined turbidimetrically.

Or determine the numbers per cm³ of the ten tubes by a microscopic count.

Then *plot* the turbidity against the tube number (= number of viable cells/cm³). A straight line should be obtained.

Use of calibration curve: This can only be used for suspensions of the named organism. i.e. separate calibration curves must be made for different pure cultures of organisms.

Sampling

In order to examine the flora of some things it is necessary to take samples. There are several techniques, all of which are easy to perform.

1. Swabbing

Samples of the flora on the surface of some objects are taken by swabbing. A swab consists of a wooden stick similar to a wooden cocktail stick around the end of which is wrapped a head of cotton wool. It is placed in a small test tube which is plugged and autoclaved (10 lb/10 minutes). It is not sterilised in the hot air oven because the cotton wool head may char a little.

Use: The swab is moistened in sterile, quarter strength, Ringer's solution (or sterile saline) contained in a sterile screw cap bottle. If a bacterial count is anticipated then it should contain exactly ten cm³ of solution. Excess fluid is removed from the swab by pressing it against the inside of the bottle. It is then wiped over the test surface, and used to inoculate suitable media. If a bacterial count is to be done after swabbing it is dropped into the Ringer's solution, the end, which has been handled, is broken off by means of sterile forceps and discarded. The lid is replaced and the swab is thoroughly shaken up in the solution to dislodge any bacteria in it and a plate count carried out using the pour plate technique. Dependent on the expected degree of contamination serial dilutions may or may not be carried out.

Swabbing crockery and cutlery: The important surfaces are swabbed:

the upper surfaces of plates,

the inside and lip of bowls,

both sides of spoons, and both sides and between the tines of forks,

the inside and the outside of cups and glasses to a depth of 3 cm below the rim.

Counts are made using 1 cm³ quantities of swab solution and plating, usually onto N.A.

Suggested experiments:

Observations on the efficiency of washing up methods by swabbing.

(i) Taking counts before and after washing, or after different methods of washing.

(ii) After a meal take 2 used forks (or moisten each with saliva), introduce one into a flask of cooled nutrient medium direct; wash up the other and introduce it into a second flask of cooled medium. Allow the media to set, incubate and observe. Gelling medium must be used.

Tracing how an organism spreads by swabbing

(i) Spread a door knob or other surface with *Serratia marcescens*. Take samples from various classrooms over the days. Look for its presence—it produces distinct red colonies on N.A. after incubation at 37°C for a day or two.

(ii) By means of insects. Allow a blowfly to crawl about on a culture of *Serratia marcescens.* Remove it and introduce it into a confined space, previously sterilised (such as a belljar). Subsequently sample the space by swabbing.

2. Water samples

Fix a handle and a long piece of string around a large sterile bottle which has a screw cap. Keep the bottle closed until you are about to obtain a sample. Remove the lid at the last possible moment and sterilise round the neck with a flame. Lower the bottle into the water in such a way that the mouth of the bottle is pointing upstream and the water is flowing into it. If there is no current create one by dragging the bottle through the water. This practice washes away any contaminating organisms from the sides of the bottle due to handling. Remove it from the water and replace the lid as soon as possible. If you can weight the outside bottom of the bottle slightly, so much the better, because this will prevent it from floating on the surface making the sample difficult to obtain.

3. Vegetation

Cut using sterile scissors and handle with sterile forceps i.e. avoid handling it with your fingers. Shake it up with ¼ strength Ringers solution and examine the solution.

4. Soil

With sterile spoons ladle soil into sterile containers. Shake up with ¼ strength Ringers and test. For counts put a weighed quantity of soil (1 gram) into a known quantity of sterile diluent (9 cm^3). This gives 1/10 dilution. Mix well and make serial dilutions. Inoculate N.A. and incubate aerobically and anaerobically at 22°C and at 37°C. (See page 59)

5. Foods

As for soil. (See also page 142). It is important with any sample with a texture like soil or foods to mix it well with the diluting

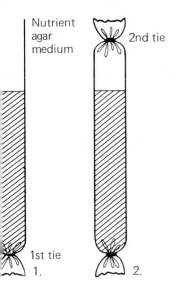

Nutrient agar medium

2nd tie

1st tie

1.

2.

— Make a *very* small hole

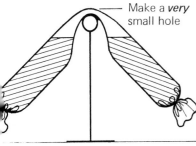

Support in autoclave

Tie here

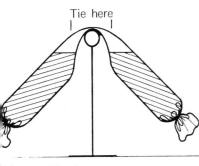

After sterilisation

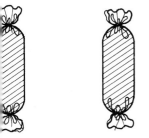

Finally—2 agar sausages—sterile.

34

fluid. Use of a sterile homogeniser greatly aids this.

6. Milk

Take samples from an unopened bottle. (See also page 129).

7. Air

Expose plates for half to one hour. (See also page 144).

Note. Do not leave samples any longer than necessary prior to inoculation of suitable media because the flora will change. Death or multiplication of the cells will take place resulting in counts which give a false picture.

8. Use of Agar sausages

A method has been evolved in recent years for sampling surfaces for bacterial contamination by using agar sausages. Melted nutrient agar media are poured into a polyamide casing to form a sausage (Ref. No. 5). This is sealed at both ends and autoclaved. Media made like this can be stored indefinitely. The casing, which must be able to withstand autoclaving, can be obtained from W.J. Kempner Ltd., 47, St. John Street, Smithfield, London E.C.1.—('Nalophan') or from Soplaril (G.B.) Ltd., 4, Watkin Road, Wembley, Middx. ('Rilsan 50/004').

Method for making agar sausages: A 65 cm length of casing of diameter 4·0 cm is tied at one end. 300 cm^3 of melted nutrient agar based medium are poured in at the other end which is then tied. It is best to use an increased % agar in the medium (2·0-2·5%). Leave several inches between the top of the medium and the position of the knot. Drape this sausage over a suitable support inside the autoclave cutting a *small* hole in the casing at the point of support and sterilise. After sterilisation allow the medium to cool down but not to set. Tie the casing firmly above the agar on either side of the support and then, when set divide into two sausages. They will each be 3·4 cm diameter and 17 cm long, and the cross sectional area will be approximately 9 square centimetres. (See Fig. 34)

Suggested media in which agar sausages can be made:

N.A.
MacConkey agar
Wort agar
Milk agar

How agar sausages are used for sampling: The outside casing of one end of the sausage is sterilised by swabbing the end with alcohol.

A broad flat knife is sterilised by flaming and then is used to cut the end off the sausage. The small piece is discarded. The medium is protruded from the casing by squeezing at the far end, and the exposed surface is pressed against the test surface. A 3-4 mm slice of the sausage is cut off and placed contaminated side uppermost

Cultures

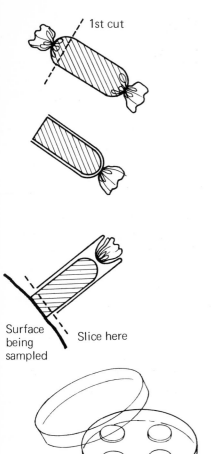

1st cut

Surface being sampled

Slice here

Four slices in a petri dish

Fig. 35
Use of agar sausages

in a sterile petri dish. The newly cut end of the sausage is ready for the second sample to be taken.

Three or four slices of the sausage can be placed in one petri dish.

If the whole sausage is not used at once it can be withdrawn by pulling out the pushed-in polyamide casing at the end and so withdrawing the medium. The casing is closed. On use it is best to cut off and discard the first slice in case it was contaminated.

Agar sausages have the advantage that they sample directly and so indicate the distribution of contaminating organisms on a surface, something which the swabbing technique does not do. (See Fig. 35).

List of suggestions for use of the agar sausages:

1. Sampling surfaces before and after washing or disinfecting.
2. In use as small 'plates'.
3. As surface for the growth of organisms from:

fingernail cuttings
sprinkling of dust
lick—from tongue and saliva
a hair
a smear from soap
a sprinkling of soil
exposed to air

5 Staining and making micro-organisms visible

Introduction

It is essential that the reality of micro-organisms should be appreciated early in their study. All work will have much more meaning if it is realised that the bacteria and other organisms are 'alive' in spite of their being invisible to the unaided eye. It is difficult for a child to appreciate that, when for example food goes bad the 'badness' is caused by the chemical activities of living creatures invisible to the unaided eye.

Experience has shown that it is insufficient to demonstrate the macroscopic effects of micro-organisms alone, and that students require to see the individual cells for themselves for, once a student is convinced of their existence his interest will be stimulated. This is possible with the aid of a good light microscope, preferably one which has an objective lens which can be used with oil immersion.

The organisms can be viewed alive or as stained smears. A visual aid which can be of great help are photographic slides of bacteria in which the magnification is high and the morphology of the cells quite clear. Such slides are sold by one firm (Banta Ltd.) and are

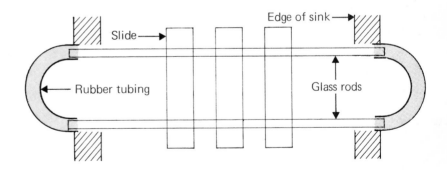

Fig. 36
Showing sink slide support

designed to be viewed through a plastic 'microscope'. They are very good. Projection microscopy of actual slides of micro-organisms, although simple, requires equipment for projected pictures which are not clear and which do not, at the moment, solve the problem of demonstrating to classes of pupils. Colour slides (photographs) for projection are available from numerous sources.

This chapter describes the ways and means of seeing micro-organisms clearly.

Preparation of the bench

For staining you will need the following equipment:

Bunsen burner. The air hole should be adjustable so that it can be open or closed. Arrange it so that you do not burn your arm when using the sink.

A rack or row of stains and staining solutions arranged in the order in which they are to be used.

A box of slides (and cover slips if required).

A dish—the base or lid of a petri dish for example—of abrasive powder.

Two or three pieces of blotting paper or filter paper.

A jam jar containing a sharpened grease pencil, a pair of forceps, a wire loop and possibly a straight wire.

The culture(s)

A sink with tap

A sink rack for supporting slides while they are being stained. A sink rack can be made of two pieces of glass rod joined at each end by a 4-5 in length of rubber tubing. The distance between the rods is just sufficient (2 in) so that they will support a 3 in slide (See Fig. 36)

Alternatively pieces of plasticene can be used to support the rods on the side of the sink—although these tend to fall off when damp.

A jar or bowl of disinfectant for containing used slides.

Microscope immersion oil and a microscope with a 1/12 in oil immersion lens.

Lawn handkerchief, for drying slides.

Slides and their cleaning and disposal

It cannot be emphasised too much that one of the secrets of making staining easy is to have absolutely grease free slides. If the slide is dirty the drop of culture, and later the staining solutions, will run into small droplets on the surface of the slide, will not form a uniform film, and will stain badly. New slides are not clean enough to be used for staining and so for them as well as for other slides the following procedure must be adopted just prior to staining.

Procedure

Moisten the finger tip with a drop of tap water and dip it in abrasive powder. Rub the slide on *both* sides. Rinse the slide and your finger under the tap until both are quite free of powder and then stand the slide in a position so that it can drain on a piece of filter paper. When it is nearly dry blot it *once* by folding the filter paper over it to form a 'sandwich' and slowly but firmly smooth the paper *once* with the side of your hand. After use, sterilise slides in the autoclave, clean them as above and store them in alcohol. Flame before use.

Test for a grease free slide

A drop of water on the clean slide will spread out in a thin film. A greasy slide will cause the water to run into droplets. Once a slide has been cleaned (that is, made grease free) keep it so. Handle it by holding the edges and then you will not leave any greasy finger prints!

Removal of grease pencil marks

These are best removed from glassware and slides with a drop or two of xylene on a paper tissue. Traces of xylene are then removed by the process described above using abrasive powder.

Disposal of slides after staining

Immediately after use slides should be dropped·into a jar full of lysol or other disinfectant placed on the bench for that purpose. When the pot is full of slides they can be cleansed and returned to general circulation. Prior to use they should be cleaned as outlined above (with abrasive powder.)

Methods for the observation of motility

It is probably easier to believe in life at a microscopic level if the micro-organisms are seen to move. Not all bacteria are motile, but those which are, move by means of organelles which are known as flagellae. (See Fig. 37)

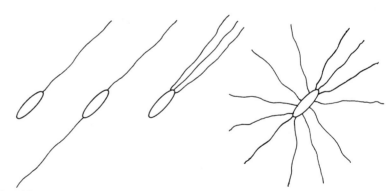

Fig. 37
Distribution of flagellae

Macroscopic demonstrations of motility

1. Using *Proteus (Proteus vulgaris):* Proteus is a non-pathogenic species which is actively motile. If one half of an N.A. or blood agar plate is inoculated and incubated at 37°C the organisms will spread beyond the area inoculated and will eventually cover the whole plate. They will not develop into discrete colonies as a non motile species would. The growth is fine and gives a crystalline appearance to the surface of the plate. (See Fig. 38)

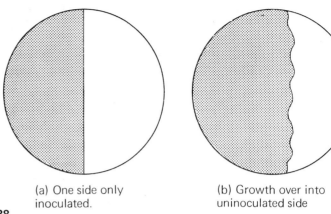

(a) One side only
inoculated.

(b) Growth over into
uninoculated side

Fig. 38

Plate showing growth of a motile
species such as *Proteus vulgaris*

2. Using a Craigie tube: In liquid media non motile organisms
and motile organisms alike are dispersed throughout the medium
by convection currents. 'Sloppy' nutrient agar is nutrient broth to
which only 0·1-0·2% agar has been added. It is a jelly like liquid
which does not allow convection currents. Thus non motile
organisms remain near their site of inoculation while motile
organisms distribute themselves throughout the medium by their
own activity.

Put 12 cm³ of sloppy agar into a universal bottle and add a
Craigie tube. Make sure that the glass tubing clears the surface of
the agar by an amount too large to allow a meniscus bridge
between the inside and the outside of the tube. Replace the screw
cap, release it half a turn, and sterilise the whole in the autoclave.
(15 lb/15 minutes.)

Inoculate the culture into the medium *inside* the tube. If the
organism is motile it should be possible, after incubation for a
suitable period, to isolate the organism by sub-culture from the
medium outside the tube. (See Fig. 39).

Microscopic demonstrations of motility

The motility of bacteria can be observed under the microscope
using a slide in one of the following three ways: using a cavity
slide, using a plain slide with cover slip and vaseline, or using a
plain slide and cover slip only.

In order to observe motility the organisms must be suspended in
fluid. Use liquid culture directly, or use the condensation water
at the bottom of a slope culture which will contain a number of
organisms. Alternatively, using a sterile pipette squirt a small
amount of sterile nutrient broth or saline over the slope and thus
suspend the organisms or put a drop of sterile nutrient broth or

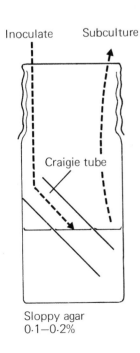

Inoculate   Subculture

Craigie tube

Sloppy agar
0·1-0·2%

Fig. 39

Showing the use of a Craigie tube
for detection of motility

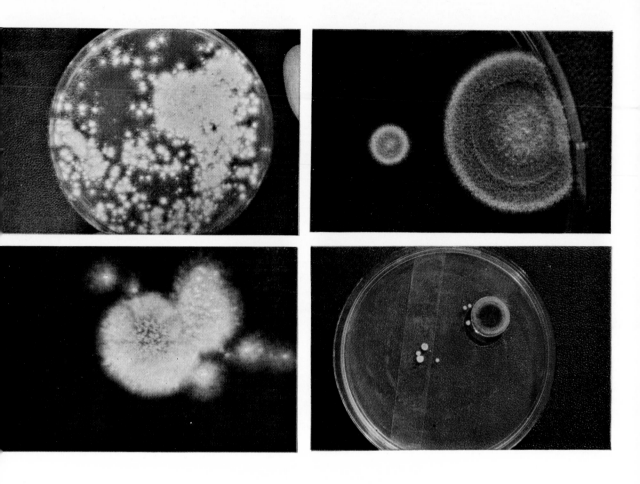

p Mould colonies on MA (⅔ life size) (c.RKP)   *Top* Mould colonies on MA (×4) (c.RKP)

ottom Mould colonies on MA (×3) (c.RKP)   *Bottom* Mould colonies on MA (Life size) (c.RKP)

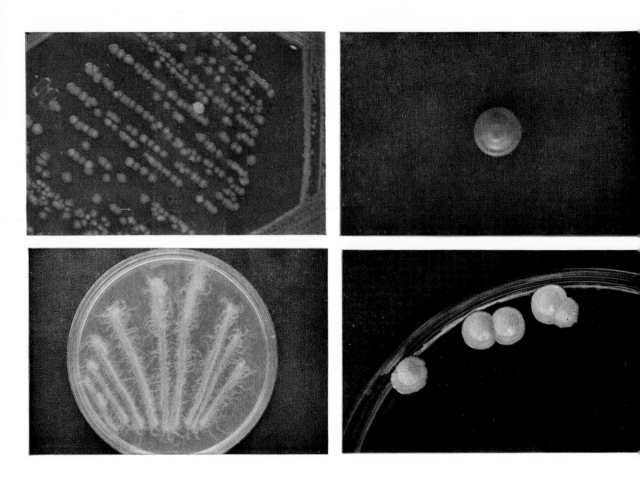

Top Bacterial colonies in NA (about half life size). The majority of these are 'entire' (c.RKP).

Bottom Bacterial colonies on NA. The colonies have run together but the edges of the streaks show 'rhizoid' growth. The organism is a species of Bacillus. (About $\frac{2}{3}$ life size.) (c.RKP)

Top A single yeast colony on MA after one week's growth. (c RKP)

Bottom Several yeast colonies on MA after one week's growth. (c.RKP)

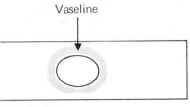

Vaseline

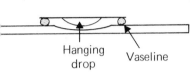

Hanging   Vaseline
drop

40

ction of motility using the
ing drop technique

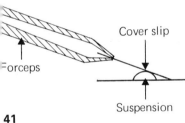

Cover slip

Forceps

Suspension

41

ering a cover slip onto a drop
ure

saline on a slide and remove a portion of a colony on a plate with a sterile loop and emulsify it in the liquid on the slide. Do not use a suspension which is very dense. A slightly opaque one is correct.

1. Using a cavity slide: By means of a match dipped into vaseline outline a ring or square round the well in the slide. With a wire loop place a drop of the prepared suspension on a clean cover slip laid on the bench. Invert the slide over the cover slip allowing the glass to adhere to the vaseline. Then quickly invert the slide so that the coverslip is uppermost. The drop should then be hanging from the coverslip into the concavity. Place the slide on the microscope stage, rack down the condenser and partly close the iris diaphragm. In low power focus the edge of the drop so that it appears across the centre of the field. Then view in high power. Do *not* use oil immersion, for if you do there will be a tendency for the cover slip to move with the oil when the field is altered. (See Fig. 40).

2. Plain slide and vaseline: Draw a square or circle of vaseline in the central portion of a plain (and clean) slide and proceed in the same way as with the cavity slide. The vaseline is very useful in preventing loss of liquid by evaporation.

3. Plain slide: Easiest of all is to put a small drop of the prepared suspension on to the clean slide and to lower the cover slip onto the drop. Take care that the culture, which is alive, is not splashed onto the bench or gets onto the hands. When judging the quantity of suspension to use the cover slip should not float, nor should there be so little that most of the space beneath the cover slip is air. In either of these two cases discard the slide and cover slip directly into a jar of disinfectant and start again.

Ideally there should be just enough liquid to 'fill' the cover slip but no more. (See Fig. 41).

Unstained organisms are difficult to see under the microscope because their refractive index is so nearly the same as the suspending fluid. The following technique enables you to focus on the living organisms quite easily.

Cut down the amount of light entering the microscope by closing the iris diaphragm. When you are lowering the cover slip in place 'plonk' it (gently) so that plenty of air bubbles are trapped. The organisms will tend to be congregated at the edge of the bubbles and it is very easy to focus on the edge of a bubble, and virtually impossible to focus on a near invisible microbe! Use the high dry objective.

There tend to be currents at the edge of bubbles so do not confuse drifting or Brownian movement with active motility. Sometimes the number of motile organisms in a culture is low compared to the number of sessile ones, so (systematically) study a number of fields, and do not conclude that a culture is non motile until you have observed several fields. You can be sure of motility if you observe an organism moving 'in a purposeful way' especially if it is moving 'against the tide'.

This is important when there is a slight drifting movement.

Disposal of slides

Dispose of slides and cover slips used for the observation of motility into a lysol pot, or pot of other disinfectant, remembering that the cultures, motile or not, are alive. Make certain that the whole slide is covered.

Examples of non pathogenic organisms which are motile.

Large bacteria (and therefore easy to see)   *Bacillus subtilis,* and many other aerobic sporing organisms.

Other bacteria, less easy to see because generally they are smaller in size   *E. coli.*
Proteus (some forms of it can be large.)
Pseudomonads.

Staining

There are three steps which are taken to make and observe a stained slide of bacteria: making a smear, the staining procedure, and possibly mounting to make a permanent preparation, and viewing.

Making a smear

A smear is a thin film of a suspension of organisms which when stained is thin enough to allow light through but is not so thin that the organisms are few and far between.

Step 1. From liquid culture: Clean the slide so that it is quite grease free.

Light the bunsen burner. Have a small flame with the air hole open.

Sterilise the loop by holding the bottom 2/3 of the wire in the flame and allowing it to become red hot. Retain red heat for about 30 seconds. Remove the loop from the flame, and allow it to cool, taking great care that the sterile loop does not touch anything.

Hold the culture tube in the left hand so that the bottom half of the tube lies against the closed fore and second fingers and is held in place with the thumb. (See Fig. 42).

Holding the loop in the right hand remove the stopper of the culture tube with the third and fourth fingers of the right hand.

Pass the top of the culture tube through the flame a couple of times, and then insert the sterile loop into the culture.

Obtain a loopful so that the film of culture stretches across the

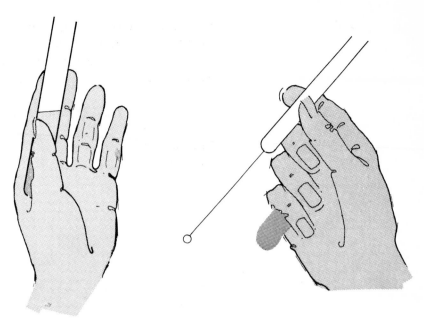

Left hand holds
culture

Right hand holds loop and stopper

Fig. 42
Manipulation of tube, its stopper
and the loop when making a smear

loop, and then remove it from the tube taking care not to touch
the sides of the test tube.

Flame the top of the test tube, replace its stopper and return it
to the rack.

Smear the loop of culture onto the slide so that it forms an even
film in the centre of the slide. It should be *just* visible as an
opaque film.

Sterilise the loop in the flame and replace it in its holder on the
bench.

Making a smear from a slope culture: Proceed as described
above either making an emulsion using a little sterile broth or using
the condensation water at the base of the slope. Take care that the
smear is not too thick.

Making a smear from a plate culture: Obtain a loopful of tap
water so that the drop hangs below the loop. Touch the clean slide
in its central portion with this in order to release the water. It
should spread out in a thin film.

Sterilise the loop and allow it to cool. Remove either a small
portion of, or a whole small, isolated colony and mix these
organisms with the water on the slide. Sterilise the loop. The final
appearance of the film should be just opaque.

Then, however you have obtained your smear, carry on as
follows.

Step 2. Drying the film: Allow the film to *dry* in its own time (a minute or two), or speed up the time, cautiously, by waving the slide with its wet side uppermost in the hot air *above* the flame. Keep the slide horizontal.

Step 3. Fixing the film: When the slide is dry, *fix* it by passing it through the flame three times. (Fixation is a histological term and is a process which kills the cells, allows them to retain a form very similar to their natural one and withstand the staining process. It also causes them to adhere to the glass). Once the film is fixed you can then stain the smear.

It is possible to make 2 or 3 smears on one slide. Take a clean, grease free slide and with a grease pencil draw thick lines to divide the middle portion into sections. A different smear can be made in each section. Make certain that sufficient length of slide is left at each side to rest on the microscope stage when viewing the end sections. (See Fig. 43).

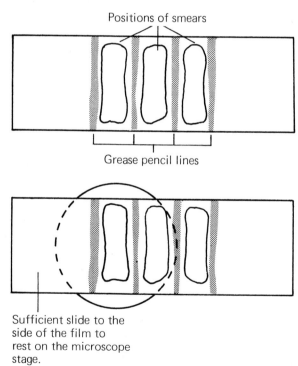

Positions of smears

Grease pencil lines

Sufficient slide to the
side of the film to
rest on the microscope
stage.

Fig. 43
Arrangement of a slide to take
more than one smear when
staining

The staining procedure

First of all check that you are going to stain the correct side of the slide. It is wise to mark the smear side with a grease pencil mark.

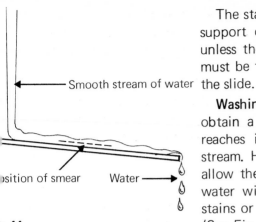

Smooth stream of water

Position of smear   Water →

Fig. 44
Washing a slide during staining

The stains are poured directly onto the slide which rests on the support over the sink. The dyes are washed off with tap water unless the staining procedure definitely requires otherwise. Care must be taken throughout the staining not to wash the smears off the slide.

Washing a slide with water, during staining: Turn the tap on and obtain a good flow of water. Slowly decrease the flow until it reaches its minimum size but is still running in a continuous stream. Hold the slide at a very slight angle to the horizontal and allow the tap water to fall onto the slide at its upper end. The water will flow over the slide in a smooth sheet washing away stains or other reagents and will not lift the smear from the slide. (See Fig. 44). Alternatively wash the slide in a beaker of clean tap water.

Drying after staining: When the staining procedure is complete the slide is washed and must be dried before it can be viewed under the microscope. Drain slides vertically onto blotting or filter paper. When all excess water has drained away blot the slide, once, with the blotting paper in the same way as has been described earlier in this chapter. Dry any remaining water from the slide by waving it in the air.

Viewing stained smears: Stained films may be observed after mounting with Canada balsam, or DePeX and a cover slip, or may be observed unmounted, but in either case oil immersion must be used. If an unmounted preparation is observed with oil, the oil can be removed later with xylol, and the smear then mounted.

Focussing using oil immersion: If a good film has been made it will be too thin to be seen easily with the naked eye and quite difficult to focus on quickly. The following method makes the process of focussing on stained bacteria very simple.

To the left or right of the film but on the same side of the slide draw a line with the grease pencil. Focus in low power on that. When it is in focus change to the next objective lens and the line should be in focus, or nearly so. When it is in focus swing to the next objective, probably the high dry, and again focus on the line. Then move that lens to one side without changing the focussing position so that neither it nor the oil immersion lens is in the optical axis. Put a small drop of oil on the grease pencil line and bring the oil immersion lens into the optical axis so that the oil joins up with the lens. The grease pencil line should be in focus.

Now carefully put a small drop of oil on the stained film, move the slide across into the optical axis, preferably using a mechanical stage. You should now find that the stained bacteria are in focus without further adjustment. This is a particularly useful method if a very thin film has been made. (See Fig. 45).

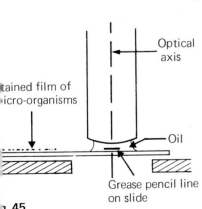

Optical axis

Stained film of micro-organisms

Oil

Grease pencil line on slide

Fig. 45
Oil immersion showing preliminary focussing on grease pencil line

Stains for bacteria

Simple stains

Methylene blue: A very simple, useful stain.

Saturday M.B.* in alcohol	30cm^3
0.01% KOH in water	100 cm^3

Method: Put a few drops of stain on the slide and leave for 3 minutes. Then wash with tap water. The time for staining is not crucial because it is difficult to overstain.

Appearance of organisms: The organisms stain blue; spores do not stain but show up as unstained spherical or ovoid areas inside otherwise stained cells. Granules in the cells may stain to a greater extent than the rest of the protoplasm giving the cell a beaded or barred appearance.

Negative staining: This is a method in which the background is stained' and the bacteria show as unstained transparent objects. It is usually used to observe the capsules of organisms, and to see spores if the spores distort the shape of the cell.

Method: A small quantity of Indian ink is placed on a slide and a small portion of a colony emulsfied in it, by mixing with a loop. Alternatively a loopful of broth culture is mixed in the ink on the slide. Then a clean cover slip is placed over the drop and pressed firmly down so that the film becomes thin and grey. Any excess is removed with a piece of blotting paper (which is discarded into disinfectant and submerged).

Appearance of the film: Capsules are seen as a clear space between the refractile cell and the grey background. In the case of spores the cells appear as refractile areas against a grey background and contain a clear area which may distort the cell shape. Some capsulate organisms: Members of the genera: *Rhizobium, Azotobacter, Bacillus, Leuconostoc.*

Gram's stain: An important stain for differentiating bacteria into two groups.
1. Gram's violet (methyl violet)
   (0.5% solution in distilled water)
2. Iodine. A weak iodine solution
   (1gm iodine, 2gm potassium iodide, 100 cm^3 distilled water)
3. 95% alcohol
4. Dilute carbol fuchsin: 1gm basic fuchsin
   100 cm^3 of solution of phenol
   (5% in water)
   10 cm^3 absolute alcohol

*Throughout the text M.B. indicates methylene blue

Dissolve the dye in the alcohol and add to the phenol.

Dilute this 1 : 10 with water to give dilute carbol fuchsin.

or Neutral red solution:

> 1 gm neutral red
> 1 litre distilled water
> 2 cm^3 1% acetic acid.

Method:

(i) Make and fix the smear.

(ii) Pour on Gram's methyl violet and leave for 30 seconds.

(iii) Pour off the excess stain. Hold the slide at an angle and add iodine so that it washes away the violet. Place the slide on the rack over the sink, add more iodine and leave it to act for about 90 seconds. Iodine acts as a mordant.

(iv) Wash off the iodine with 95% alcohol. Hold the slide at a slight angle, pour on the alcohol at the upper end and watch it push' the Gram's violet in front of it down the slide. Add more alcohol till no more violet comes out of the upper end of the smear. If you wait until absolutely no more violet appears from anywhere you will severely over decolorise.

(v) Wash immediately with running tap water, remembering the danger that the film may be washed from the slide if you are not sufficiently careful.

(vi) Apply the counterstain for 2-4 minutes. (Less than this is required when using dilute carbol fuchsin.)

Appearance of the film: Gram positive bacteria resist decoloration, retain the methyl violet and therefore appear purple;

Gram negative bacteria do not resist decoloration, lose the violet stain and on counter staining are coloured pink.

There is a fundamental difference in the structure of the cell wall of these two types, and the Grams reaction is a basic characteristic used in the taxonomy of bacteria.

Gram positive organisms: Genera: Staphylococcus, Streptococcus, Lactobacillus, Clostridium, Bacillus.

Some gram negative organisms: Genera: Escherichia, Proteus, Salmonella, Shigella, Pseudomonas.

Stain for spores (Dorner's spore stain): Spores resist normal staining processes and so methods to stain them positively are fairly drastic.

Strong carbolfuchsin (refer to Gram's stain for recipe)

¼-½% H_2SO_4

1% aqueous methylene blue.

Method:

(i) Stain with strong carbol fuchsin for 3—5 minutes, heating the preparation until steam rises. (Use a torch of cotton wool wound round a piece of strong wire, dipped in alcohol, and ignited. Wave this beneath the slide on its support over the sink. *Or* place over one corner of a tripod, and with a burner heat a second corner). Do not boil the stain nor allow it to run dry. As the stain evaporates replenish it.

(ii) Wash with water.

(iii) Treat with the sulphuric acid (1-3 minutes), until the film takes on a yellow colour. This time varies with the organism which is being stained, and it may be found necessary to use a stronger acid.

(iv) Wash in water.

(v) Counterstain with methylene blue (3 minutes).

(vi) Wash, drain and dry.

Appearance of the film: Spores are stained bright red, protoplasm blue.

Some sporing organisms: The aerobic sporing organisms, Genus *Bacillus.* The anaerobic sporing organisms, Genus *Clostridium.* Samples of both of these can easily be isolated from soil cultures.

Simple mounts and stains for fungi
Lactophenol blue

Phenol crystals	20gm
Lactic acid	20cm³
Distilled water	20cm³
Glycerol	40cm³
Cotton blue or methyl blue	0·075gm

Dissolve the phenol crystals by gentle warming, then add the dye. **Caution: Phenol is highly corrosive and must not be handled.** Add the other reagents.

Method: Place a drop of the stain on the slide and in this gently tease a fragment of the culture with needles. Apply the cover slip with some pressure as far as possible eliminating bubbles. Remove excess stain with blotting paper. This stain takes time to penetrate and is best after several hours. The edges can be sealed with cellulose lacquer or clear nail varnish.

Glycerol: Fungi can be mounted in a drop of glycerol or other wetting agent, and observed.

Both of the above methods are for temporary mounts only.

'Cellotape' method: Place a drop of lactophenol cotton blue on a slide. Impress the sticky side of a 3–4'' length of 'Cellotape' over a mould colony and then stick it lengthways over the slide. Leave 5–10 minutes. Observe under the high power lens. This is a good method for observing moulds because it minimises disturbance of their structures.

Slide culture of moulds: With a sterile scalpel cut a 3mm square of malt agar from a malt agar plate, and place aseptically on a sterile slide. Cover the square with a sterile cover slip. With a loop inoculate the sides of the square with mould spores. Place the culture in a petri dish; cover and incubate at about 25°C overnight. Observe the slide under low power—germinating spores should be visible.

6 Some elementary practical exercises

The object of this chapter is to enable the inexperienced to practise techniques and gain proficiency in micro-biological methods.

Preparatory work

Assemble these items for personal use:
forceps,
slides,
coverslips,
distilled water,
immersion oil,
jar of lysol,
grease pencil.

Experiment 1. Making small items of equipment

Make the following items: (See Chapter 1. page 12)
loops,
glass spreaders,
mouthpieces,
pasteur pipettes.

Experiment 2. Preparation and cleaning of glassware

Prepare any glassware you require for the following sequence of experiments for example:
test tubes,
flasks,
petri dishes and so on. (See Chapter 1. page 9)

Experiment 3. Making media

Make and sterilise the media you require for subsequent experiments. (See Chapter 2, page 21)

Testing the efficiency of sterilisation methods
Experiment 4.

To show that steaming at 100°C for 90 minutes is an efficient sterilisation method

Principle: Organisms are knowingly introduced into a tube of media in which they can grow. Some of the tubes are sterilised, some are not. If the sterilisation method is adequate no growth will appear in the sterilised media while it will appear in the unsterilised tube.

Timing for experiment

Day 1: Make *one* broth. Label it *tube A.* Do not sterilise it. Leave it exposed to the air and wait for visible growth to appear. It can be incubated either at room temperature or in the incubator at 37°C.

Day 2 or later, (when growth has appeared). Make a series of 9 nutrient broths, about 5 cm³ each. With a sterile loop (flaming) add an equal quantity of culture from test tube *A* to 6 of the broths, tubes *B*1, *B*2, *B*3, *B*4, *B*5, *B*6. The inoculum should be small enough not to make any visible difference to the clearness of the medium. Do *not* inoculate tubes *B*7, *B*8 and *B*9. Give all nine

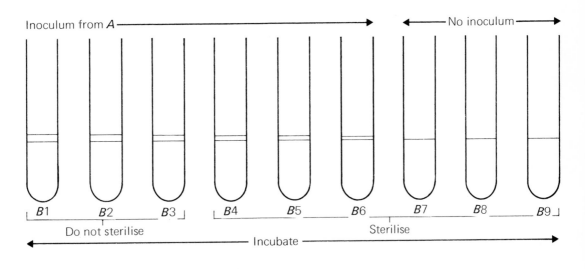

Fig. 46
Arrangement of tubes to show that steaming at 100 °C for 90 minutes is an efficient method of sterilisation

tubes a metal cap or cotton wool stopper (See page 12). Place tubes *B*4, *B*5, *B*6, *B*7, *B*8, *B*9 in the steamer and steam for 90 minutes. Do not steam *B*1, *B*2, *B*3,.

Incubate all nine tubes under the same conditions as tube *A.* (See Fig. 46).

Results: If sterilisation is adequate no growth should appear in tubes *B*4 to *B*9. Growth should appear in *B*1, *B*2 and *B*3.

Experiment 5
To show that autoclaving at 15 lb for 15 minutes (or 10 lb for 30 minutes) is an efficient method of sterilisation

This experiment is conducted exactly as Experiment 4 except that the sterilisation is done using an autoclave and not a steamer.

Experiment 6
To show that Tyndallisation is an efficient method of sterilisation

Principle: Organisms are inoculated into media in which they are known to grow. Some of the inoculated media are not sterilised at all; some are sterilised once (30 minutes at 100°C), some are sterilised twice (30 minutes at 100°C twice on successive days) and some are given the full Tyndallisation procedure. Viable organisms may be present but not visible in liquid media immediately, but after several days if any are present growth should appear.

Method: Make 6 N.A. for slopes in screw cap bottles, sterilise them by a proven method (autoclave or steamer) and then make slopes. These can be prepared several days in advance of the experiment.

Timing for the experiment:

Day one: Make one N.B. Do not sterilise it. Expose it to the air, (or use a culture of a known sporing organism for example, *B. subtilis*, or inoculate the N.B. with a little soil) and allow visible growth to appear. Label it tube *A*.

Day two: Make 12 N.B. in test tubes. Label them *B*1, *B*2, *B*3, *C*1, *C*2, *C*3, *D*1, *D*2, *D*3, *E*1, *E*2, *E*3. With a sterile loop inoculate all 12 tubes using '*A*' as the culture. Each tube should have a metal cap or cotton wool stopper. Place *B*1, *B*2, *B*3 in the incubator without sterilisation. Place all *C, D* and *E* tubes in the steamer for 30 minutes. Incubate all broths.

Day three: Treat broths *D*1, *D*2, *D*3, *E*1, *E*2, *E*3, for a further 30 minutes in the steamer. When cool inoculate slopes from each of the *D* tubes, label the slopes and incubate all the broths and slopes overnight.

Day four: Sterilise broths *E*1, *E*2, *E*3 for a further 30 minutes in the steamer. When cool inoculate slopes from each tube. Label the slopes and incubate all broths and slopes overnight.

Day five: Assess all results. Examine broths *C*1, *C*2, *C*3, slopes *D*1, *D*2, *D*3, *E*1, *E*2, *E*3 for growth. Tyndallisation procedure gives spores present a chance to germinate. Broths which have not had the full treatment are likely to show growth since spores not killed will have the chance to germinate. Since, in the case of *D* and *E* tubes they may be killed by subsequent treatment, viable organisms must be detected by inoculation of slopes (dead cultures are still opaque). The slopes inoculated from the tubes receiving the full treatment the *E* slopes should not show growth. (See Fig. 47).

Experiment 7
To show that sterilisation using hot air is an efficient method of sterilisation

Media known to be sterile (sterilised by a proven method, see Experiments 4, 5 or 6) are introduced aseptically into glassware which has undergone sterilisation in the hot air oven. Any growth in the media after sterilisation is attributed to organisms present in the glassware.

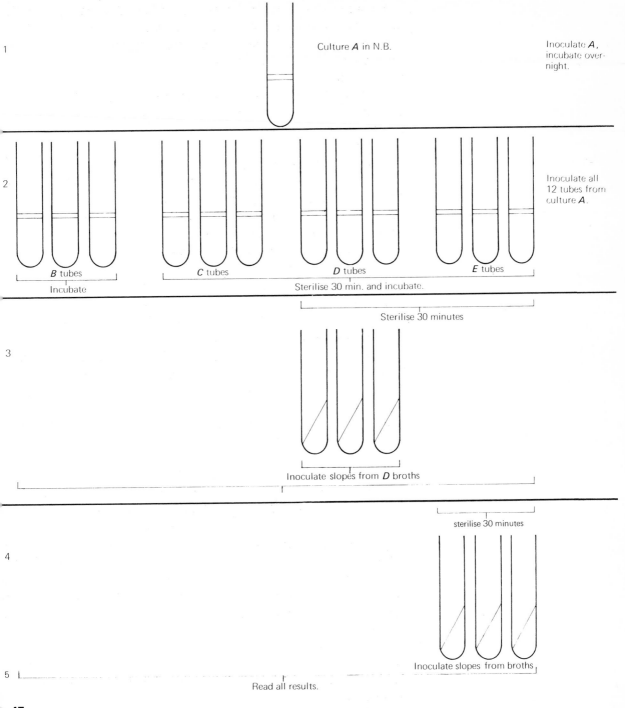

1   Culture *A* in N.B.   Inoculate *A*, incubate overnight.

2   Inoculate all 12 tubes from culture *A*.

B tubes   *C* tubes   *D* tubes   *E* tubes

Incubate   Sterilise 30 min. and incubate.

Sterilise 30 minutes

3

Inoculate slopes from *D* broths

sterilise 30 minutes

4

Inoculate slopes from broths

5   Read all results.

g. 47
rk plan to show that Tyndal-
tion is an efficient method of
rilisation

Method: Make 6 N.B. in test tubes (about 5 cm³ each. Sterilise these by a proven procedure). Each tube should have a cap or stopper. Label: *A*1, *A*2, *A*3, *A*4, *A*5, *A*6. Sterilise three test tubes (empty, dry, chemically clean, stoppered) in the hot air oven

(160°C for 60 minutes). Sterilise 3, 10 cm³ pipettes (Refer to page 46).

Leave the media untouched in tubes *A*1, *A*2, *A*3. Using a separate sterile pipette on each occasion aseptically transfer the sterile broth from *A*4, *A*5, *A*6 to the test tubes. Label these *B*4, *B*5, *B*6. Incubate *A*1, *A*2, *A*3, *B*4, *B*5, *B*6. No growth should develop in any tube. (See Fig. 48).

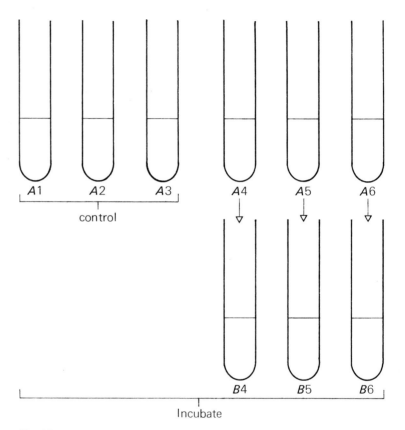

Fig. 48
Arrangement of tubes to show
that use of hot air is an efficient
sterilisation method

Experiment 8

1. **Sterilise tubes** containing filter paper discs of culture in hot air, and subsequently incubate the discs in nutrient media. Growth or no-growth will indicate the adequacy of the treatment.

Experiment 9

2. **Use of spore strips.** Spore strips are pieces of paper which are impregnated with the spores of a very heat resistant organism— very often *Bacillus stearothermophilus.* These have to be purchased.

Load the hot air oven in the normal way with equipment to be sterilised and secrete spore strips in various places in the oven. After sterilisation the strips are aseptically removed and dropped into separate tubes of N.B.—which are then incubated. Since *Bacillus stearothermophilus* is a thermophilic (heat loving) organism it is better to incubate it at a high temperature—55-65°C if this is possible. If sterilisation has been adequate there should be no growth in any tube. Growth in some tubes and not others indicates uneven heat distribution in the oven, more often than not due to overfilling it.

Experiment 10
To show that boiling water kills vegetative organisms, and can therefore be used to sterilise equipment

Principle: Organisms are knowingly introduced into water which is afterwards boiled. Samples of this water are inoculated into sterile nutrient media and incubated. Growth in these media would indicate that sterilisation had not taken place.

Method:

Day 1: Make ten N.B. and sterilise nine of them by a proven method. Leave the tenth one open to the air for 30 minutes or so and then keep it until visible growth appears (culture *A*)—or use a broth culture of a known organism.

Day 2: Inoculate N.B. No. 1 from tube *A*. This acts as a control. Take a sterile stoppered flat based 250 cm^3 round flask. Put in it about 50 cm^3 of tap water. Boil the water. Then carefully tip into the boiling water the culture *A*. Boil. At timed intervals remove a loopful using a sterilised loop and inoculate in turn the nine broths. (Replace flask stopper after each manoeuvre). Incubate all nine broths.

Day 3: Assess all results. Growth should appear in some tubes. The pattern of growth will depend on the time of exposure to boiling water and the strain and type of organism used. (Boiling water coagulates the proteins of the cells—a parallel can be drawn with boiling eggs.) (See Fig. 49).

Vegetative cells are much more readily destroyed than spores.

Experiment 11
To demonstrate the effect of washing on the microbial flora of the hands

Method: Pour three *large* plates. These can be made using the top of a 7lb biscuit tin, a sheet of glass acting as the lid. 50-60 cm^3 of N.A. will be required for each plate. One hand is deliberately dirtied by rubbing it over the bench top or floor and then an imprint of that hand is made on plate 1. The same hand is then elaborately and very thoroughly washed with soap ensuring that a good lather is obtained. The hand is then well rinsed in running water and shaken to become dry (do *not* use a towel). The damp,

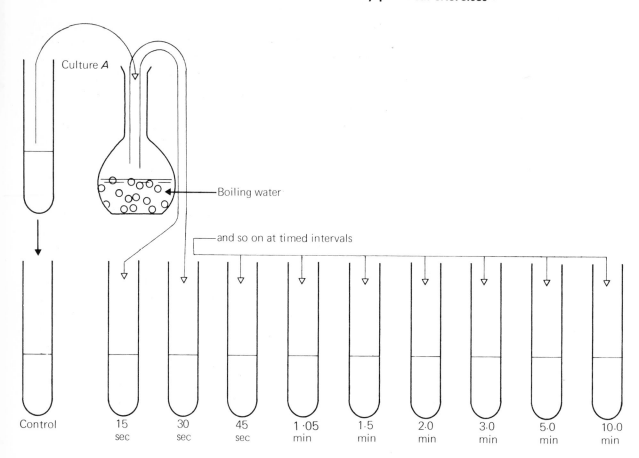

Culture *A*

Boiling water

and so on at timed intervals

| Control | 15 sec | 30 sec | 45 sec | 1·05 min | 1·5 min | 2·0 min | 3·0 min | 5·0 min | 10·0 min |

Fig. 49
Plan of work to demonstrate that boiling water kills vegetative organisms

clean hand is then imprinted on plate 2. Dry hands on a clean paper towel and imprint the dry hand on plate 3.

Plates 1, 2 and 3 are incubated overnight, and a good demonstration result should follow.

Note that washing reduces the variety of micro-organisms on the hands, but that the skin saprophytes (*Staphylococcus albus*) abound after washing. This experiment illustrates the value of frequent hand-washing in the reduction of cross contamination, but also shows that the hands can be a source of organisms— sometimes they can be a source of the food poisoning organism *Staphylococcus aureus.*

Working with organisms
A series of experiments could be set on the following lines:
  1. Exposure of plates to air, soil, water and so on.
  2. Isolation of species, and inoculation of plates by various methods. (See Experiment 12).
  3. Making stock cultures of these species.
The techniques necessary to carry out these experiments are all covered in detail in Part I of the book.

Experiment 12

Isolation of species, and inoculation of plates by various methods

Method: Examine the plates which have been exposed to air, soil, water etc. and observe the different species as identified by colonial size, shape, elevation and edge. The shine, opacity and pigmentation also help in distinguishing the different types of organisms growing on a plate. There will be both very obvious differences (and more subtle ones) which are learned with experience.

Select two or three obviously different colonies and sub-culture from these into broth.

At the same time as you sub-culture the colonies examine them microscopically for motility, and their morphology by staining.

When the cultures have grown in N.B. repeat all that you did in the previous step and check that the results are the same.

Sub-culture from the broth onto nutrient plates and inoculate to get isolated colonies. (Practice the different methods.) Incubate the plates and after incubation check that the plate cultures are pure, that is, of only one colonial type. Also examine the plates critically and see whether you are satisfied with your cultures from the technique point of view. Are they spread well? Have you ploughed up the agar? If you are satisfied with your cultures and their purity you can now proceed and make stock cultures of the selected organisms. (See Chart 10)

Chart 10

Colonial identification

Bacterial colonies are described using the following criteria, and these points are used to distinguish separate types in mixed culture.

1. Colony size—applied only to well isolated colonies. Colonies growing very close together do not achieve a characteristic size due to overcrowding, depletion of nutrients, accumulation of toxic end products of metabolism slowing down growth rates. Individual species may achieve different size colonies on different media.

2. Shape of colony—edge characteristics: entire, undulate, rhizoid, serrate,
Elevation above the medium: flat, raised, convex, umbonate, papillate.

3. Pigmentation—none—where no pigment is produced colonies have a pale beige colour due to the large number of cells present.
colours: white, orange, golden, yellow, red, pink, green.

4. Surface—dull, shiny, granular.

5. Colony texture—mucoid, spreading, dry.

Shapes   *Elevation*

entire   flat

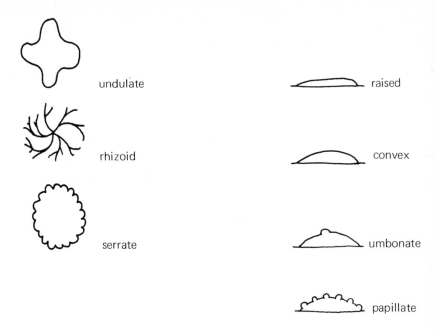

undulate

rhizoid

serrate

raised

convex

umbonate

papillate

7 Antimicrobial substances

Antimicrobial substances are used to kill or prevent multiplication of micro-organisms. Broadly speaking they are applied in two circumstances: inside or in contact with, human or animal tissues . . . antibiotics and chemotherapeutic substances; and in contact with an inanimate background . . . disinfectants.

In both circumstances the antimicrobial substance must act without damage to the background substance, obviously a more delicate and selective process when the background itself is living tissue.

The efficacy of both antibiotics and disinfectants depends on the combined effect of a number of factors: the exposure time, the concentration, the temperature, the pH, and the presence of inactivating organic material.

Generally the longer the exposure time and the more concentrated the disinfectant the greater the disinfectant action. This is not a linear relationship; twice the concentration may for example bring about 10 times the effect.

Some definitions may be useful:

Disinfectants are used against an inanimate background and divide into two groups:

Those conferring loss of viability possess the suffix '—cidal' e.g. germicidal substances. Those conferring a temporary loss of reproductive capacity possess the suffix '—static' substances. e.g. 'bacteriostatic'—active against bacteria.

Chemotherapeutic substances selectively kill micro-organisms and may be produced artificially e.g. sulphonamide drugs.

Antibiotics are substances produced by micro-organisms which are antimicrobial in action. Although many have been recognised relatively few can be used medicinally because only few are toxic to micro-organisms and not to mammalian tissue that is, a few are chemotherapeutic.

Antimicrobial substances exert their effect in different ways:

1. By coagulation of proteins. Some substances throw the colloidal particles which constitute protoplasm out of solution e.g. H^+, Cu^{++}, Zn^{++}, Fe^{+++}, Hg^{++}, Ag^+, and other heavy metallic ions. They are usually applied as solutions of their salts, such as $CuSO_4$, $AgNO_3$ and $HgCl_2$. Other substances act in the same way: alcohol, phenol (carbolic acid) formaldehyde and related and derived substances. Thus vital enzymes in the cell membrane and within the cell are destroyed.

2. Chemically active substances whose destructive action is non-specific such as chlorine, bromine, iodine, phenol, strong acid, strong alkali.

3. Specific chemical action. These substances act in ways whic depend on chemical differences between microbial an mammalian cell structure such as some antibiotics.

4. Microbes can be killed in *non-specific ways* e.g. dryin raising the osmotic pressure and so on.

Overall antimicrobial agents affect transport into and out of th cell, affect and disrupt vital metabolic reactions; affect the DNA structure and hence impair or alter its function.

The following series of experiments illustrates two points:
Substances which have antimicrobial effects.
The principles of disinfection.

1. Substances which have antimicrobial effects

Experiment 13

To show that various heavy metals present in coins have antimicrobial effects

Materials: Various well scrubbed coins: pennies or halfpennies

Well dried N.A. plates (numbers depends on how extensive the experiment is to be)
1 24 hour broth of *Staph albus* (Gram positive)
1 24 hour broth *E. coli*. (Gram negative)
2 sterile pasteur pipettes
2 sterile glass spreaders or 1 loop.

Method: With a sterile pasteur pipette and teat mix the firs culture well. Remove about 0·5 cm³ of it and drop it onto the of the plate. Spread the inoculum with the spreader or loop so that it covers the whole plate and confluent growth will result Then firmly place three or four coins on the surface of the plate as in the diagram. Incubate preferably at 37°C for 24-48 hours Repeat for the other culture.

Results: (See Fig. 50): Inhibition of growth round coins wil occur if a high enough concentration of the heavy metals present in them leaches into the medium.

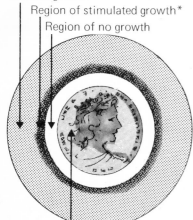

Normal growth
Region of stimulated growth*
Region of no growth

Fig. 50   Coin
Lawn plate showing the effect that heavy metals in a coin have on the growth of a micro-organism

*Sometimes bactericidal and bacteriostatic substances stimulate growth in low concentrations, whilst inhibiting at higher concentrations.

Experiment 14 (i)

To show that the salts of heavy metals have antimicrobial effect using crystals

Materials: Small crystals of various heavy metal salts:
$CuSO_4$, $AgNO_3$, $FeCl_3$
A number of N.A. plates (well dried) equal to the number of salts to be tried
1 24 hour broth culture of *Staph. albus* (Gram pos
1 24 hour broth culture of *E. coli* (Gram negative)
2 sterile pasteur pipettes
Sterile glass spreaders or loops

Method: With a grease pencil divide the back of the plate into quarters. Then with a sterile pasteur pipette and teat mix first culture well. Remove about 0·5 cm^3 of it and drop it onto the surface of the plate. Spread the inoculum with the spreader or the loop so that it covers the whole plate and confluent growth occurs. Then firmly place a *small* crystal of one of the test substances in the centre of one of the segments of the plate. Place three other test crystals in the other three segments. Incubate at 37°C for 24-48 hours this time with the agar side downwards (so that the crystal does not drop off.) Repeat for the other culture.

Results: Refer to Experiment 15.

Experiment 14(ii)
To show that the salts of heavy metals have antimicrobial effects using impregnated paper discs or strips

Materials: A selection of strips of filter paper (or blotting paper) soaked in solutions of salts of heavy metals. Other materials as for the previous experiment (14(i)).

Method: As for the previous experiment. Place the strip or disc on the seeded plate. Press it firmly down in place so that the whole of it is in contact with the surface of the agar. Incubate.

Results: See Experiment 15.

Experiment 15
To show that other substances have an antimicrobial effect

Materials: As for the previous two experiments. Soak strips or discs of suitable paper in phenol, iso-propyl alcohol, lysol, dettol, soap solution, formaldehyde, bleaching solution, dilute solutions of proprietary disinfectants and so on.

Method: As for the previous two experiments (14(i) and 14(ii))

Results for experiments **14(i), 14(ii)** and **15**: inhibition of growth will occur round the antimicrobial substance. There will be a concentration gradient of it in the medium and at the point at which its concentration is sufficiently low to be tolerated by the test organisms growth of that organism will occur. No valid conclusions can be drawn by comparing the antimicrobial substances used from the size of zone of inhibition since the rates of diffusion may differ. The only valid conclusion is whether a substance is inhibitory or not to a particular organism in those circumstances.

Experiment 16
To show the action of antibiotics

Materials: Well dried N.A. plates.
24 hour broth of *Staph albus* (Gram positive)

24 hour broth of *E. coli* (Gram negative)
Sterile pasteur pipettes
Sterile glass spreaders or loops
Antibiotic impregnated papers
Oxoid Multodiscs.

Method: With a sterile pipette and teat mix the first culture well. Remove about 0·5 cm^3 of it and drop it onto the surface of the plate. Spread the inoculum with the spreader or loop so that it covers the whole plate and confluent growth results. Then firmly place the antibiotic impregnated papers onto the N.A. and press well down. Incubate 24-48 hours. Repeat for the other culture.

Results: Sensitivity to any particular antibiotic is indicated by a region of no growth around the antibiotic disc.

2. Principles of disinfection

Experiment 17
To demonstrate that disinfection depends on the length of exposure time
Materials: Test tubes containing one of the following (or any other) disinfectant.

1/1000 mercuric chloride
1/100 iso-propyl alcohol
dilute iodine soln. (about 7 p.p.m. available iodine)
dilute phenol, lysol, or proprietary disinfectant
1/10 000 gentian violet
For each disinfectant four test tubes of sterile N.B.
N.B. culture of *Staph. albus*
Sterile pasteur pipettes, or 0·1 cm^3 pipettes
Loop
Stop clock

Principle: Organisms are exposed to the action of a selected disinfectant by inoculation of the disinfectant with a sample of the organisms. At regular time intervals a small sample of those organisms are withdrawn with a loop and inoculated into nutrient broth, which is then incubated. Some will show growth and others will not. Where no growth occurs the time of exposure to the disinfectant is enough to kill all organisms; where growth is shown the exposure time is not sufficient to kill all organisms at that particular concentration of disinfectant.

Method 1: Using one disinfectant only—such as 1/1000 mercuric chloride: The key to conducting this experiment successfully is organisation. Mark the N.B. tubes 1, 2, 3, 4 and record that tube 1 will receive an inoculum after 2½ minutes exposure to the disinfectant, tube 2 after 5 minutes, and so on. Inoculate the tube of disinfectant with 0·1 cm³ of the culture organisms using a sterile pipette, or 1 drop from a sterile pasteur pipette, and set the stop clock going at once. After exactly 2½ minutes, with a *loop* remove a sample and inoculate N.B. number 1. After another 2½ minutes (5 minutes from the start) remove another loopful and inoculate tube number 2, repeat at 7½ and 10 minutes. Incubate the four tubes for 18-24 hours at room temperature, or at 37°C.

Method 2: To compare the effects of different disinfectants. Materials as method 1 suitably duplicated to suit the number of disinfectants to be used.

The method is the same as method 1 except that several disinfectants are used at the same time.

Label each tube of nutrient broth both with the time of exposure to disinfectant that the inoculum received, *and* the disinfectant used. Timing becomes complicated. For each disinfectant timing begins at the time of its inoculation.

For example samples may be removed from disinfectant *A* at times 0, 2½ 5, 7½, and 10 minutes, but for disinfectant *B* at times ½, 3, 5½, 8, 10½ and so on. See table below:

	Disinfectants	*A*	*B*	*C*	*D*	*E*
Total exposure time to disinfectant	Times, in minutes, to inoculate disinfectants with culture	0	½	1	1½	2
2½ 5 7½ 10	Sub-culture into N.B. at	2½ 5 7½ 10	3 5½ 8 10½	3½ 6 8½ 11	4 6½ 9 11½	4½ 7 9½ 12

Results for method 1 and 2

Exposure Time	Disinfectant				
	A	*B*	*C*	*D*	*E*
2½ minutes	+	+	+	+	+
5 minutes	+	+	±	+	−
7½ minutes	±	+	−	+	−
10 minutes	−	±	−	±	−

The results should show that a longer exposure time to a

disinfectant will bring about a greater degree of disinfection.

+ = growth. Insufficient exposure time to have a bactericidal effe

± = slight growth—a mainly bactericidal effect. One or two slightl
more resistant cells survived the agent and are now reproduc-
ing.

− = no growth—a total bactericidal effect. No viable cell introduc
into nutrient birth.

Experiment 18
To demonstrate that disinfection depends on the concentration o an antimicrobial substance

Materials: A series of test tubes containing dilutions of
a chosen disinfectant such as crystal violet in
the following dilutions: 1/1 000 000, 1/800 000,
1/600 000, 1/400 000, 1/200 000,
1/100 000; or dilutions of proprietary disinfectant
For each dilution one N.B.
24 hour broth culture *Staph. albus*, or other
selected organism
Sterile pasteur pipette or 0·1 cm^3 pipette
Loop
Stop clock

Principle: Organisms are exposed to the action of a chosen disinfectant in serial dilution for a standard length of time (5 minutes). Then a tube of N.B. is inoculated with a loop and incubated to see whether or not growth will occur. If growth occurs exposure for the standard length of time at a particular concentration is not sufficient to bring about disinfection. If growth does not occur then that exposure time to that particular concentration is sufficient to bring about disinfection.

Method: Mark the nutrient broth tubes 1/100 000, 1/200 000 and so on, to indicate from which tubes of disinfectant the inoculum has originated.
Introduce into the 1/100 000 tube of disinfectant 0·1 cm^3 of culture. Set stop clock going. Introduce into the 1/200 000 tube 0·1 cm^3 of culture and note its time of inoculation. Inoculate the third and other tubes of disinfectant, and note their times of inoculation. At exactly 5 minutes remove a sample of the first tube using a loop and inoculate the first broth. At exactly 5 minutes after its inoculation remove a sample from the second tube and inoculate the second broth and so on, so that the organisms at each concentration are exposed for a standard length of time 5 minutes. Incubate at 37°C for 24-48 hours.
Result: See Experiment 19

Experiment 19

To demonstrate the combined effect of the concentration and exposure time of an antimicrobial substance on the growth of micro-organisms

Materials: As for Experiment 18 except for each dilution of disinfectant 4 tubes of nutrient broth will be needed.

Method: Experiment 18 can simply be modified. The nutrient broth tubes are labelled with *both* the exposure time *and* the dilution of disinfectant to which the organisms are subjected. The procedure is the same as before. Samples are removed from a dilution of disinfectant at regular intervals say 2½, 5, 7½ and 10 minutes from the time of inoculation. Incubate for 48-72 hours.

Results for Experiment 18 and 19: Record in tabular form whether or not growth has occurred. Absence of growth indicates that the antimicrobial agent had a bactericidal effect. This experiment can be used to find out the correct dilution of disinfectant to use against specific target organisms i.e. it is a 'use-dilution' test.

Experiment 20

To show the susceptibility of different organisms to disinfectants at graded concentrations

Method: The previous experiment can be repeated using a selection of organisms (instead of just one) or using a non-spore former such as *Staph. albus* and a sporing organism such as a five day old culture of *Bacillus subtilis.* Should a less complicated experiment be required gradient plate techniques can be used. (See later Experiment 24).

Result: See Experiments 18 and 19—Bacterial spores are extremely difficult to destroy chemically and this experiment can therefore be used to demonstrate that chemicals can be effective for disinfection but cannot be relied on to sterilise.

Experiment 21

To show the effect of temperature on disinfection

Method: Repeat Experiment 18 at each of the following temperatures: ice, room temperature, 37°C and at 60°C (in a water bath). Hold the tubes of disinfectant at the correct temperature prior to inoculation so that each can equilibrate. Inoculate with organisms and sub-culture from them at the correct time intervals into broths. Incubate the broths at 37°C for 48-72 hours. (Note it is the temperature of disinfection that counts not incubation temperature).

Result: See Experiment 19.

3. Experiments with antibiotics

Experiments which demonstrate the action of antibiotics and also antibiotic assay methods; and experiments to show classical advances in the history of medicine associated with antibiotics.

Experiment 22

To demonstrate the effect of antibiotics on various micro-organisms

Materials:	Well dried N.A. plates
	24 hour broth cultures:

        E. coli

        Staph. albus

        5 day broth culture *B. subtilis*

        Sterile pasteur pipettes

        Sterile glass spreaders, or loop

        Selection of antibiotic impreg-

        nated papers*

If the cultures are very thick dilute them to 1/10 in N.B.

Method: With a sterile pipette and teat mix the first culture. Remove about 0·5 cm³ of it and drop it onto the surface of the plate. Spread this inoculum with the loop or sterile spreader so that it covers the whole plate and confluent growth will result. Remove any excess culture with the pipette. Then firmly place antibiotic papers onto the N.A. and press well down. Incubate for 24-48 hours. Repeat for the other cultures. Note. *B.subtilis* forms rather stringy growth which may be difficult to mix.

Results: Where an organism is sensitive to a particular antibiotic a zone of inhibition round the antibiotic paper will appear after incubation.

Experiment 23

To show the action of antibiotics by cutting troughs in the N.A. and filling these with liquid antibiotic

Materials:	Well dried N.A. plates
	24 hour broth cultures:
	Staph. albus
	E. coli
	Sterile pasteur pipettes
	Sterile glass spreaders or loop
	Antibiotic solutions
	Scalpel
	1·0 cm cork borer
	1 MacCartney bottle of sterile melted N.A.

Method: With a sterile pipette mix the first culture. Remove about 0·5 cm³ of it and drop it onto the surface of the plate. Spread the inoculum with the spreader or loop so that confluent

---

*Oxoid Multodiscs, or antibiotic discs, or homemade strips or circles of absorbent paper soaked in antibiotic

growth results. Leave the plate for at least 15 minutes while the fluid part of the inoculum sinks into the medium. Then either cut *holes* or *troughs:*

To cut troughs sterilise the scalpel in the flame. Make a rectangular incision in the agar, remove the piece and dispose of it.

To cut holes sterilise the cork borer and press it firmly into the agar in the places required, but do not put the holes too close together. Five on a plate is about maximum. Remove the pieces of agar (a sterile dissecting needle can be of use).

When the holes or troughs are cut they must be sealed. This is to prevent the antibiotic from running underneath the entire piece of agar in the plate. With a fine sterile pasteur pipette drop a small quantity of sterile melted N.A. into the trough or holes to seal the edges. (See Fig. 51). Dispose of the pipette into hot or boiling water. Then with a sterile pipette (a different one for each antibiotic solution) very carefully place a drop of antibiotic in the trough. Repeat for the other culture. Incubate for 24 hours at 37°C.

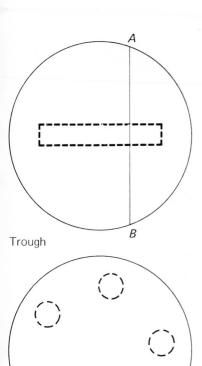

Trough

Holes

g. 51

ıtting and sealing holes made in A. plates

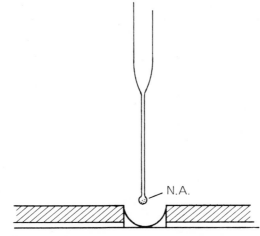

To seal a hole or trough

Results: The antibiotics will diffuse into the agar. There will be a concentration gradient round the trough. The growth of sensitive organisms will be inhibited and so there will be a zone of no growth round the trough. This is a qualitative test and so zone size cannot be taken as indicative of relative sensitivity.

Experiment 24
To demonstrate the effect that different concentrations of antibiotic (or disinfectant) have on different organisms using the gradient plate technique

Materials: 1 sterile petri dish
10 cm³ melted N.A.

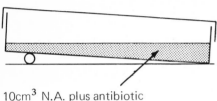

10cm³ N.A. plus antibiotic
10cm³ N.A.

A   *B*

10cm³ N.A. plus antibiotic

Fig. 52
Gradient plates

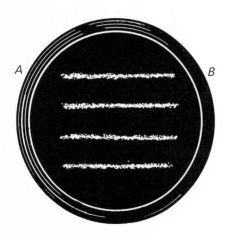

Inocula parallel to the
concentration gradient *AB*.

Fig. 53
Parallel inocula on a gradient plate

10 cm³ sterile N.A. plus antibiotic
24 hour broth cultures of:
Staph. aureus
Staph. albus
E. coli
Pseudomonas species
Loop

Principle: Nutrient agar containing antibiotic is layered beneath the nutrient agar in a plate in such a way that a concentration gradient of antibiotic is formed. Parallel streaks of organisms are inoculated parallel to the gradient. Organisms which are sensitive to the antibiotic grow in the regions of the lowest concentration only, those which are less sensitive have longer streaks of growth on the plate. (See Fig. 52).

Method: Pour 10 cm³ of nutrient agar containing the antibiotic into the petri dish and allow it to set in a tilted position so that on one side of the plate the agar is deep, and on the other shallow. When this layer is quite set remove the support from beneath the plate, and place the plate flat on the bench. Pour on the N.A. and allow it to set. With a loop inoculate the plate with the first organisms in a single straight line parallel to the concentration gradient, and streak out the other organisms parallel to this. Incubate overnight at 37°C. (See Fig. 53).

Results: See 'principle' above.

Antibiotic Assay Methods

The term 'assay' means to find out how much antibiotic is contained in a given fluid. The principle of assay methods is to set up a series of dilutions of the 'unknown' solution and to inoculate each with a standard volume of culture of a known sensitive organism. This organism will have a tolerance of the antibiotic (i.e. will be able to grow), up to a certain concentration. In concentrations greater than that the organism will be inhibited and will not be able to grow. The lowest concentration which inhibits the growth of the organisms is known as the M.I.C. (Minimum Inhibitory Concentration). If the sensitivity of an organism is known then the actual value of the M.I.C. is also known, and so the 'unknown' fluid has been assayed.

Experiment 25
To assay an antibiotic solution by the tube dilution method
Materials: 10 sterile plugged test tubes
                      1 24 hour broth of a sensitive organism
                      such as *Staph. aureus*
                      1 tube sterile antibiotic
                      10 sterile 5 cm³ pipettes

60 cm³ sterile N.B. in a flask

1 sterile pasteur pipette.

Principle: Using a pasteur pipette a standard inoculum of a 24 hour broth culture of an organism is added to a series of N.B. containing doubling dilutions of antibiotic. A control tube of nutrient broth is also inoculated and observed for growth. The first tube in which no visible growth is observed gives the titre, or sensitivity, of the organisms to the antibiotic. (See Fig. 54) (Note. In contrast to disinfectants the organisms are incubated in contact with the antibiotic.)

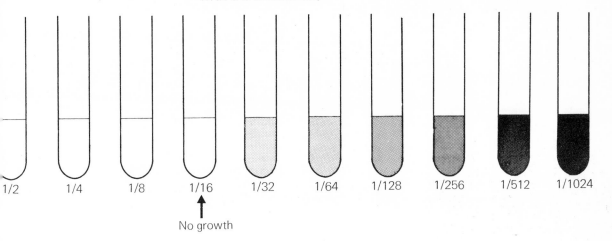

| 1/2 | 1/4 | 1/8 | 1/16 | 1/32 | 1/64 | 1/128 | 1/256 | 1/512 | 1/1024 |

↑
No growth

nsitivity of the test organism is
16 x strength of neat antibiotic

ppose the sensitivity of the test
ganism is known to be 5 units/
³

is inhibited by the antibiotic at
16 dilution

. 54
he dilution for the assay
antibiotic or for finding out
e sensitivity of an organism

∴ 1/16 tube represents a concentration of 5 units/cm³ (approx.).
∴ strength of the neat antibiotic
= 5 x 16 = 80 units/cm³ (approx.)

Method: Arrange the sterile tubes in a rack and label them ½, ¼, $\frac{1}{8}$, $\frac{1}{16}$ and so on. With a sterile pipette put 10 cm³ N.B. into the control tube, and 5 cm³ into the dilution tubes. Then using the sterile pipettes make doubling dilutions of the antibiotic in the broths. When the series of dilutions has been made mix the culture well and withdraw a sample in the sterile pasteur pipette. Holding it absolutely vertical introduce one drop of culture into each of the tubes including the control. Discard the pipette into disinfectant or into a boiling water bath. Incubate all tubes at 37°C for 24 hours.

Results: After incubation some tubes will show growth and some will not. (See Fig. 54).

This experiment can be used in two ways:

(1) If the strength of neat antibiotic is known the sensitivity of the organism can be computed from the results

(2) if the sensitivity of the organism is known the actual strength of the neat antibiotic can be calculated.

Experiment 26

To assay an antibiotic using nutrient agar plates

Materials: 5 tubes melted N.A. (15 cm³ exactly)
5 sterile petri dishes
sterile pipettes
24 hour broth cultures:
Staph. aureus
Staph. albus
E. coli
Loop
tubes of antibiotic

Principle: Small quantities of antibiotic are introduced into the melted agar and well mixed to ensure uniform distribution throughout. The agar is poured and allowed to set and parallel streaks of the test organisms are inoculated on the surface of the plates. This method can either be used to compare the reaction of different organisms to varied strengths of antibiotic or to assay fluids.

Method: Calculate the quantities of antibiotic to be added to the nutrient agar to give the final strength desired. Arrange these to give doubling dilutions, plate to plate. It is important to remember that the antibiotic will dilute the agar, so care must be taken not to add too much, which would prevent the agar from gelling. Remove an equal volume of agar before the antibiotic is added. Proceed thus: Label the tubes of melted N.A. with the dilution of antibiotic that each will contain. With a sterile pipette remove from each the calculated quantity of agar, and replace each tube in a hot water bath to prevent the agar from gelling. To each in turn add the appropriate quantity of antibiotic. Mix well. (Alternatively put the antibiotic in the petri dish, add the melted N.A. and mix the two by tilting and rotating the plate.) Pour the plate and allow to set. Then streak inoculate with test organisms. Incubate 37°C for 24 hours.

Results: (1) For any organism its M.I.C. is the lowest concentration of antibiotic which inhibits growth. This is the sensitivity of the organism.

(2) By knowing the sensitivity of the organisms used the antibiotic containing fluid can be roughly assayed. Suppose $\frac{1}{16}$ plate prevents growth of an organism whose sensitivity is 5 units/cm³

Therefore concentration of plate ($\frac{1}{16}$) is at least 5 units/cm³
The strength of neat antibiotic = 80 units/cm³.

Experiment 27

To examine the potency of an antibiotic by the trough method.

Materials: Well dried N.A. plates (one per organism).
24 hour broth cultures:
Staph. albus
E. coli
Sterile pasteur pipettes.
Doubling dilution of antibiotic solutions.
Sterile glass spreaders or loop.
1·0 cm cork borer.
1 MacCartney bottle of sterile melted agar.

Method: Proceed exactly as for Experiment 23. Put one drop of each dilution of antibiotic into the sealed holes in the plate. Incubate each plate for 24 hours at 37°C.

Results: In this case the size of the zone of inhibition can be taken to indicate the potency of different strengths of the antibiotic on a plate. But unless the experiment is very carefully standardised with respect to test organism, strength and consistency of agar, time and temperature of incubation and pH, and so on, comparisons cannot be made from one plate to another, or from one experiment to another.

Experiment 28

To examine the potency of an antibiotic using impregnated discs

Materials: Nutrient agar plates
24 hour broth cultures of
Staph. albus.
Staph. aureus
E. coli
Sterile pasteur pipettes
Sterile glass spreaders or loop
Antibiotic impregnated discs (graded concentrations
· of one antibiotic)

Method: Approximately 0·1 cm³ of the first culture is dropped onto the surface of an agar plate using a sterile pasteur pipette and is spread to give confluent growth. The plate is left for about 15 minutes while the liquid sinks into the agar. Then impregnated discs containing measured amounts of antibiotic are pressed onto the agar in much the same way as in previous experiments. Incubate 37°C for 24 hours.

Results: Interpretation of these results is subject to the same provisos as are the results of Experiment 27.

Note: This method is used to assay the potency of a new antibiotic preparation, to determine how much antibiotic there is in the body fluid of a patient on chemotherapeutic drugs i.e. an 'unknown' antibiotic solution and for testing the susceptibility of organisms to antibiotic.

Experiment 29
Antibiotic production by moulds etc. and their inhibitory effect on strains of organisms

    Materials: Mould culture such as *Penicillium* species
                  Flask containing 250 cm^3 Sabouraud dextrose medium.

    Method: Inoculate the flask of medium with the mould and incubate at room temperature for several days until good growth is obtained.   Separate the mould from the medium by centrifuging or by filtration through sterile filter paper into sterile containers.

    Results: This filtrate will contain any antibiotic that has been produced by the culture and can be tested for and assayed in the ways described previously.

Experiment 30
To demonstrate the presence of antibiotics in food stuffs

    Antibiotics are known to act as growth stimulants and are sometimes added to the foods of animals. They are sometimes used to 'water' crops as a control against plant pathogens. Their use is under strict legislation but nevertheless traces of antibiotics may appear in the foods we eat. Their presence in milk and foods may be tested for as follows:

    Sterile absorbent paper discs are dipped into the test milk, or in emulsified food and placed on prepared lawns of culture and incubated.

    The presence of antibiotic will be shown by zones of inhibition round particular discs.

Use as test organisms:
Staph. albus
Sarcina lutea

Note: This rough method will only give demonstrable results if the antibiotic is present in high enough concentration and if the organism used is sufficiently sensitive. For an accurate method to estimate penicillin in milk refer to: *'Bacteriological techniques for dairy purposes'. Technical Bulletin no. 17, Min. Ag. Fish. Food. Pub. by H.M.S.O. page 103.*

Experiment 31
To show antibiotic production in the soil

    Materials: Soil
                  1 nutrient broth—sterile
                  1 sterile petri dish
                  2 10 cm^3 N.A.—melted cooled

24 hour broth of *Staph. albus*
Loop or spreader
Pasteur pipette—sterile

Method: Make a suspension of soil in N.B., and dilute it 1/10 (approximately). Add about 1 cm^3 of this to 10 cm^3 sterile, melted cooled N.A. Pour the whole in the sterile petri dish and allow it to set. Pour 10 cm^3 melted, cooled N.A. over the culture and allow it to set. Incubate the culture 37°C for a short while (4-6 hours). Make a lawn of the *Staph.* on the surface of the plate and withdraw any excess fluid. Incubate at 37°C for 24 hours.

Results: Any antibiotic produced by the organisms in the deep layer will diffuse through to the top layer and produce zones of inhibition in the top lawn.

A similar experiment could be performed using known organisms to see whether they produce any antibiotic substances.

A series of experiments to illustrate various stages in the history of antibiotics

(1) Asepsis and aseptic surgery. Lister 1827-1912.
Carbolic acid spray—(phenol). The sensitivity of organisms to various strengths of phenol can be demonstrated using lawn plates and impregnated discs. The degree of contamination of surfaces can be demonstrated as in the section on *sampling.* (See page 73).

(2) Discovery of penicillin—Fleming 1929. Demonstration of production of antibiotics—Experiment 29.

(3) Discovery of other antibiotics—produced by soil organisms—Waksman—Experiment 31.

(4) The problem of the development of resistant strains (See Chapter 8. 'Genetic Methods').

8 Genetic methods

Introduction

Bacteria possess chromatinic material which appears to act as a single chromosome. Usually bacteria multiply by simple binary fission. The frequency with which it occurs varies according to both the species and the circumstances. Fission can, in at least one species, be as frequent as once every 20 minutes resulting in very large numbers of bacteria in a short time. Since a lot of generations are produced in this short time bacteria are very suitable for experimental work in genetics and inheritance.

The 1000 or so genes on the chromosome of bacteria are all capable of mutating. These changes in the genetic constitution of the bacterium are brought about in different ways:

transduction by 'phage—that is pieces of genetic material are transferred from one bacterium to another by a bacterial virus known as bacteriophage.

conjugation of two bacteria—physical union of 2 organisms facilitates transference of DNA from one to other.

transformation—experimental extraction and addition of bacterial DNA.

spontaneous mutation

induced mutation.

Spontaneous mutations

Many reasons have been put forward to explain the fact that genes on the bacterial chromosome mutate spontaneously, including the possibility that cosmic rays may sometimes be instrumental. Some of the genes have a tendency to mutate faster than others. The fastest mutate at a rate of about once in every 10^6 cell divisions, the slowest at a rate of about once in about every 10^{11} or 10^{12} cell divisions. In practice, this means that there will be at least one mutant of a preselected type in a bacterial population of 10^{12} organisms. Thus, provided that suitable selection techniques are used, such mutants can be identified.

Induced mutations

Various chemicals and some wavelengths of the electromagnetic spectrum are used to induce a faster rate of mutation. Most of the subjected culture will be killed by the mutagenic agent but in the surviving cells the mutation rate for any particular gene will be found to have increased. It is thought that the high number of cell deaths is partly due to a number of mutations which are lethal. Examples of mutagenic agents are ultra-violet and infra red light, picric acid, nitrogen mustard. Gene mutations show themselves in various ways:

change in morphological properties

metabolic loss or gain of enzymes
change in resistance to antibiotics
change in the degree of pathogenicity

Prototrophs

These are the wild type of organism, that is the type with which the mutants are compared.

Auxotrophs

Auxotrophs, are bacterial mutants which exhibit a loss in ability to synthesise an essential metabolite, due to a gene mutation which has resulted in the loss of ability to synthesise an essential enzyme. In order to grow and multiply, therefore, auxotrophs must be supplied with the essential metabolite in the medium.

The problem in experimental work with bacteria is to be able to demonstrate the existence of the mutants. The following methods have been found to be of great use:

the replica plating technique
observation of morphological changes: pigmentation
                                        colonial form
                                        sectoring of colonies

Replica Plating Technique

The principle of this technique is that an imprint giving an exact replica of a culture plate is transferred to several plates each containing different media. This means that the growth characteristics of all the colonies on the original plate can be studied simultaneously.

A stamp is made from a circular piece of well seasoned, close grained wood and of a size slightly smaller than a petri dish. The wood should be at least 3 cm thick. Cut a circular piece of best velvet, for the better the quality the better the stamp will work. Use 'Copydex' latex adhesive to stick the velvet onto the wood block with the pile facing away from the wood. (Elek and Hilson recommend using the best velour, 2·5 mm Ref. 9). This must be sterilised so wrap it in Kraft paper and sterilise in the hot air oven. The stamp can be used again and again, and is only discarded when the pile begins to lie flat.

Experiment 32

To detect antibiotic resistant mutants of a strain of *Staph. aureus* normally sensitive to the antibiotic

Materials: 1 24 hour broth culture of *Staph. aureus*
            [Use Experiment 22 page 106 to select a
            suitable antibiotic]
            1 N.A. plate
            5 or 6 N.A. plates containing graded quantities
            of an antibiotic such as penicillin
            1 sterile wrapped velvet stamp
            Loop

Method: Spread the N.A. plate with *Staph. aureus*, using the loop, in such a way to obtain isolated colonies over the whole plate. Incubate this plate overnight at 37°C.

Label each of the antibiotic plates with their concentration and also with a reference mark. Put a line on the stamp as a corresponding reference position. After incubation take the N.A. plate culture and press the stamp firmly down on it, taking particular care not to smudge the plate. Then align the reference marks and stamp antibiotic plate 1 again taking care not to smudge; proceed to plate 2, align the reference marks and stamp the plate, and so on for all antibiotic plates.

Keep the N.A. plate, incubate the antibiotic plates at 37°C overnight.

Results: There will be growth on plates at low concentrations of the antibiotic and less growth on the higher concentrations. By careful comparison of plates you will be able to identify colonies on the original N.A. plate which contain mutants to the antibiotic. These are spontaneously occuring mutants that is, they have not been induced by the presence of the antibiotic because you can see them on the ordinary N.A. plate. (See Fig. 55)

Experiment 33
To show the presence of spontaneously occuring mutants to different antibiotics

Materials: 1 24 hour broth *Staph. aureus* (whose sensitivity to antibiotics is known. Experiment 25).
3 N.A. plates
Antibiotic discs
1 sterile wrapped velvet pad
Loop

Method: Spread the 2 N.A. plates with the test organism in such a way as to obtain confluent growth (plates 1 and 2). Allow the plates to dry. Then carefully press onto the surface of plate 2 a selection of antibiotic discs. Incubate both plates overnight at 37°C. Make a reference point on the velvet pad and on plates 2 and 3. Press the pad onto plate 2 and imprint onto plate 3, having first aligned the reference points. Incubate plate 3 at 37°C for 24 hours, and keep the other two plates.

Results:

Plate 1: This is the control. There should be confluent growth over the whole plate.

Plate 2: There should be confluent growth over the whole plate except in circular zones around the antibiotics to which the organism is resistant.

Plate 3: The replica plate. It is expected that there will be one or two colonies growing inside the 'clear' zones of plate 2. These

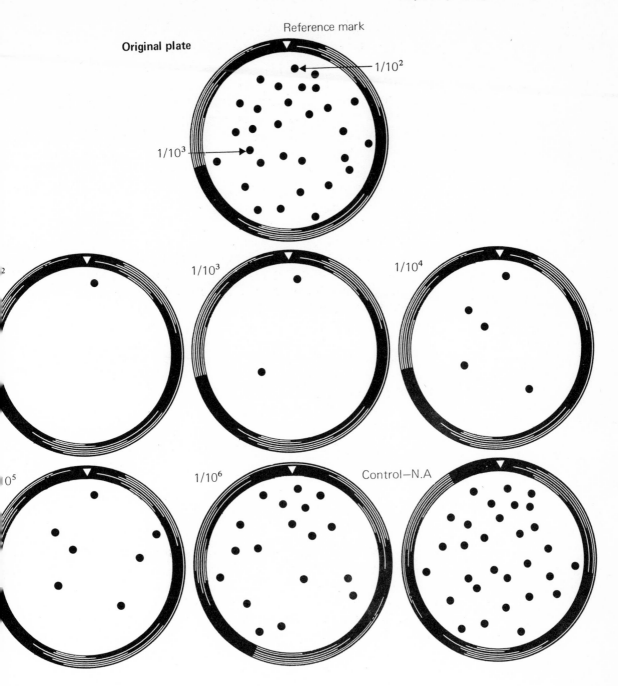

Original plate

Reference mark

1/10²

1/10³

1/10²

1/10³

1/10⁴

1/10⁵

1/10⁶

Control—N.A

Fig. 55
Antibiotic resistant mutants
detected by the replica plate
technique

will be colonies of the staphylococcus derived from cells not killed by the antibiotic, and which are therefore more resistant than the parent culture (See Fig. 56). These are antibiotic resistant mutants.

Fig. 56
Possible results for experiment 33

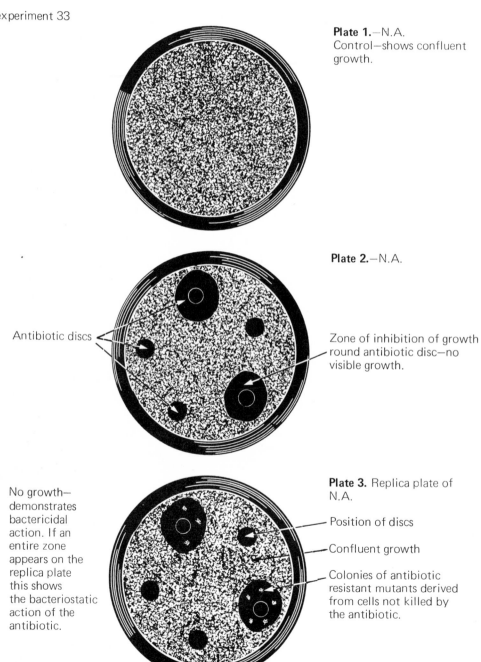

Plate 1.—N.A.
Control—shows confluent growth.

Plate 2.—N.A.

Antibiotic discs

Zone of inhibition of growth round antibiotic disc—no visible growth.

No growth—demonstrates bactericidal action. If an entire zone appears on the replica plate this shows the bacteriostatic action of the antibiotic.

Plate 3. Replica plate of N.A.

Position of discs

Confluent growth

Colonies of antibiotic resistant mutants derived from cells not killed by the antibiotic.

This experiment can both be used to demonstrate the bacteriostatic and the bactericidal action of the antibiotic. Looking at the results obtained on the replica plate we see:

(1) no growth on the sites of inhibition round the antibiotic discs indicates bactericidal action,

(2) growth on the sites of inhibition of the antibiotic discs indicates bacteriostasis.

This experiment could be followed by investigating just how resistant the mutants are, possibly by gradient plate methods.

Gradient plates
Experiment 34
To develop resistant strains of a micro-organism

Materials: 10 cm³ N.A.
10 cm³ N.A. plus antibiotic
Sterile petri dish
24 hour broth culture of *Staph. aureus*
Loop

Method: Make a gradient plate of the N.A. and the N.A. plus antibiotic. Streak inoculate the organism onto the plate parallel to the gradient. Incubate 37°C for 24 hours. Growth of the pattern shown in Fig. 57(a) will be found. Spread the colonies growing in the higher concentration region in the way shown in Fig. 57(b). Incubate at 37°C for 24 hours. Repeat this procedure, if necessary until growth is obtained in the region of highest antibiotic concentration. Then resuspend resistant colonies in nutrient broth and streak out in a similar manner on a gradient plate with 2-5 times higher antibiotic concentration in the antibiotic.

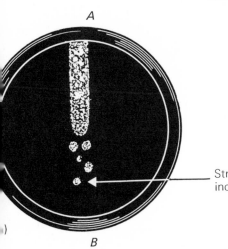

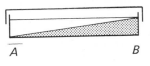

Streak out this colony and re-incubate.

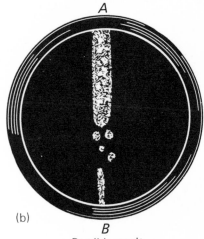

(b)

B

Possible result
after incubation

g. 57
:velopment of resistant strains
micro-organisms

Experiment 35

To measure the comparative resistance of mutants

Materials: Gradient plate, containing antibiotic
              Broth cultures of mutants
              Loop

Method: Streak the mutants over the plate parallel to the concentration gradient. Length of growth is a direct measure of resistance (See Fig. 58).

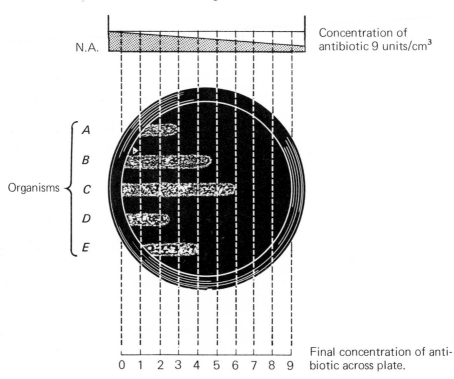

Fig. 58
To measure the comparative
resistance of mutants to antibiotic

Metabolic mutants

Sometimes changes in the metabolism of organisms in a colony show by affecting the colonial appearance. When this happens it is a useful means for demonstrating the occurrence of spontaneous mutation.

Escherichia coli normally ferments lactose producing acid. If neutral red is incorporated into the solid lactose containing medium the lactose fermenting colonies will be bright pink due to the production of acid during fermentation. Any mutation which occurs during growth and which affects the ability to ferment lactose will show as a change in the intensity of the pink of the affected colony. The easiest way to detect such colonies is to look

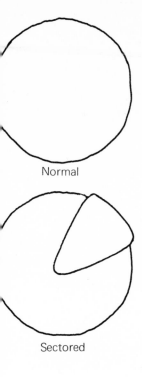

Normal

Sectored

Fig. 59

Sectored colony

for sectored colonies. These show a section, usually a quarter or a well defined segment of a colour different to the rest of the colony. This indicates that a mutation has occurred during the early development of the colony giving rise to a clone of cells of different fermentative ability (See Fig. 59).

Experiment 36
To demonstrate lactose fermenting mutants of *E. coli* by sectored colonies. (A good class experiment).
    Materials: Culture of *E. coli*
                 12 MacConkey agar plates
                 Loop
    Method: Spread the plates with the culture in order to obtain isolated colonies. Incubate the plates overnight at 37°C. Study each plate carefully for sectored colonies. Do not expect to find many. There may be one or two in the whole 12 plates. (See Fig. 59).

These simple genetic experiments may be used as they are or modified to illustrate:
(1) Fluctuations in character of natural communities.
(2) Demonstration of natural selection.
(3) That characters such as the pathogenicity of some organisms change giving rise to new strains.
(4) The care that must be exercised in the therapeutic and other uses of antibiotics.

9 Examination of water

Introduction

Water, being ubiquitous, always contains a certain flora of organisms, the variety of which depends on from where the water is derived. Deep well water, and fast flowing mountain water are very often practically sterile; slow moving water usually contains a number of organisms, as do sea water, pond water, ditch water, and soil water. The type of organisms present depends on a number of factors: the type of soil over which the water flows, whether it is open to contamination by animals, whether human sewage and industrial effluents are passed into it and whether it is subjected to any kind of purification.

Generally speaking all different types of water can be sampled and examined for their flora in ways already described (See page 48 Chapter 4).

The more specific methods described in this chapter are based on those used in the bacteriological examination of water to ensure that it is pure and fit for human consumption. The examination, which takes place before it reaches our homes, is for the presence of certain types of bacteria, whose presence indicates pollution by faeces. Organisms which are pathogenic to, and may be carried by man and animals may reach water via their faeces, but tend not to survive very long since their growth requirements are fairly exacting. Since water is supplied to all homes and buildings it can act as the vehicle for the spread of disease—especially of the typhoid and dysentery types, and it is therefore of the utmost importance that water is examined regularly, and any sign which indicates the possibility of pollution must be acted upon.

If pathogenic organisms do get into water it is probable that their numbers will be low and so to prevent the job of the micro-biologist being like looking for a needle in a haystack techniques have been developed whose results may indicate contamination of water supplies and therefore the possible presence of pathogens. If drinking water is shown to be contaminated by faeces, human or animal, steps will be taken immediately to prevent further contamination. Any type of water, tap, pond, ditch, river, sea, lake, reservoir etc., can be examined by the following methods.

Taking samples

See 'Water samples' page 74 chapter 4.

Examination for organisms

The pathogenic organisms of the dysentery and typhoid groups are delicate and survive and thrive in the gut of human beings and animals. The gut is also heavily populated with many thousands of millions of commensals, which are always present in faeces. These are of the family Enterobacteraceae. These commensal organisms are generally speaking hardier than their pathogenic relations, and so it is for these commensal faecal organisms that water is examined. If they are found to be present, then it is possible that the pathogens are also present but in lower and probably in undetectable numbers.

Characteristics of the Enterobacteraceae

They are Gram negative rods

Motile, or non motile

Non-sporing

Ferment glucose

Some are intestinal pathogens: typhoid, paratyphoid, food poisoning and dysentery organisms (Salmonellae and Shigellae).

Some are intestinal commensals: such as *E. coli, Proteus*

Some are saprophytes in the soil and water, such as Aerogenes group

Since many of the Enterobacteraceae live in the intestine, they are able to tolerate the presence of bile in their environment, something which very few other organisms are able to do. Water is therefore examined using bile-containing media to discourage the growth of other organisms.

Experiment 37

To demonstrate whether or not Enterobacteraceae are present in water samples

Materials: Water samples in sterile bottles

For each water sample to be tested:

3 tubes 10-15 cm³ MacConkey agar, single strength, melted, cooled

3 sterile petri dishes

1 x 1 cm³ pipette sterile, with mouthpiece

Method: Mix the sample well by rotating the sample bottle between the hands. Into each petri dish pipette 1 cm³ of the sample water. Carefully add a tube of melted cooled MacConkey agar to each plate, making certain that the whole plate is covered with medium. Incubate the plates at 37°C for 24-48 hours.

Result: Colonies on or in the medium will probably be those of Enterobacteraceae. They may be rose pink due to lactose fermentation or colourless if they do not ferment lactose. The organisms should be gram negative rods.

This rough method only indicates whether or not members of the Enterobacteraceae are present and does not indicate their origin which could be faecal, or could be from soil or from both.

Experiment 38
To carry out the presumptive coliform test

This is a test which shows the presence of coliform organisms (the gut commensals) and indicates the probable numbers present in the original sample.

Materials: 50 cm³ double strength MacConkey broth in a 200 cm³ flask and containing a Durham's tube, sterile
5 test tubes with 10 cm³ Double strength MacConkey broth with Durham's tubes, sterile
10 test tubes containing 5 cm³ single strength MacConkey broth
Sterile pipettes: 6 10 cm³
5 1 cm³
5 0·1 cm³

Method: Follow Fig. 60. All tubes are incubated at 37°C for 18-24 hours after which they are examined. All tubes showing acid, a change in the medium to a bright pink, and an amount of gas sufficient to fill the concavity at the top of the Durham's tube are regarded as 'presumptive positives' that is, they are presumed to contain coliform organisms. A lesser volume of gas than this may be disregarded, unless visible gas appears in the liquid when the tubes are lightly tapped. Positive tubes are submitted for further examination, negative tubes are incubated for a further 24 hours. If any tubes are seen to be positive after this second period of incubation then these too are submitted to further tests.

Results: From the numbers of tubes showing acid and gas at 37°C the probable number of coliform organisms present in the original sample of water can be estimated. (See tables in the appendix, and also Reference 7).

The further test to which positive tubes are subjected are those which differentiate the types of coliform organisms present and which indicate if any *E. coli* type I (indicative of human faecal pollution) are present. If type I is found to be present it means that the water has definitely been polluted with faeces, and could therefore contain human pathogens.

Experiment 39
The differential coliform test

From all tubes in the presumptive coliform test which are positive inoculate a single strength MacConkey broth and incubate at 44°C for 24 hours. *E. coli* type I lives only in the human intestine. It can produce acid and gas at 44°C, something which the saprophytic Enterobacteraceae cannot do. Therefore positive

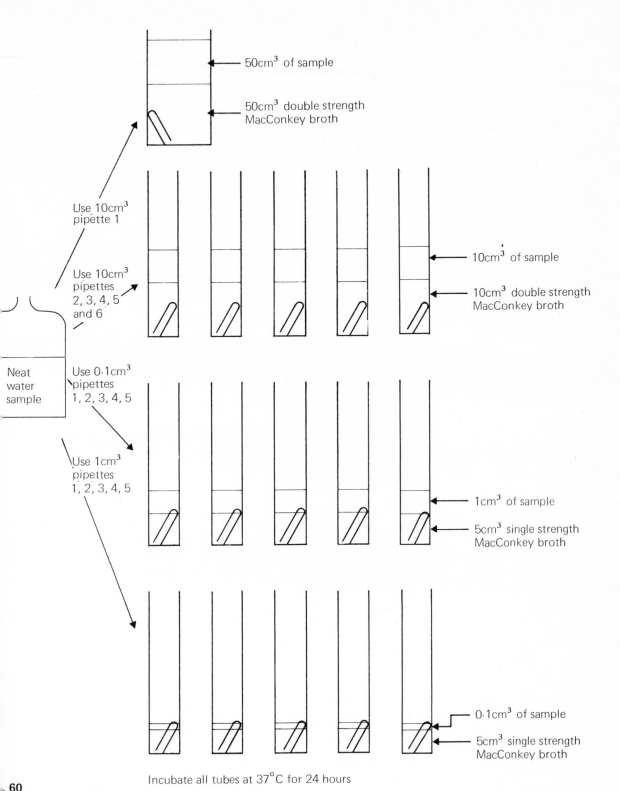

50cm³ of sample

50cm³ double strength MacConkey broth

Use 10cm³ pipette 1

Use 10cm³ pipettes 2, 3, 4, 5 and 6

10cm³ of sample

10cm³ double strength MacConkey broth

Neat water sample

Use 0.1cm³ pipettes 1, 2, 3, 4, 5

Use 1cm³ pipettes 1, 2, 3, 4, 5

1cm³ of sample

5cm³ single strength MacConkey broth

0.1cm³ of sample

5cm³ single strength MacConkey broth

Incubate all tubes at 37°C for 24 hours

sumptive coliform test

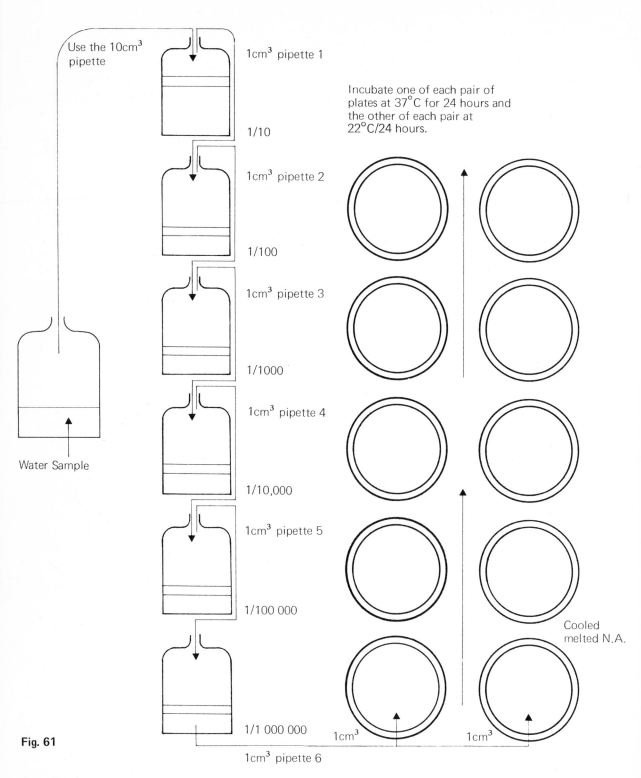

Use the 10cm³ pipette

1cm³ pipette 1

1/10

Incubate one of each pair of plates at 37°C for 24 hours and the other of each pair at 22°C/24 hours.

1cm³ pipette 2

1/100

1cm³ pipette 3

1/1000

Water Sample

1cm³ pipette 4

1/10,000

1cm³ pipette 5

1/100 000

Cooled melted N.A.

1/1 000 000

1cm³   1cm³

Fig. 61

1cm³ pipette 6

Counting the number of organisms
in a sample of water

tubes for this test indicate the presence of type I.

To double check this result all positive tubes are tested to see whether the coliforms present can produce indole.

Indole test: Sub-culture from all positive tubes into peptone water (see page 23). Incubate at 37°C for 24 hours. Add Ehrlichs Rosindole reagent. A rose colour indicates the presence of indole and confirms the presence of *E. coli* type I., proving the faecal contamination of the original sample.

(This experiment can be 'set up' by inoculating samples with *E. coli* type I).

Experiment 40

To count accurately the number of organisms in a sample of water

Materials: 1 flask containing the water sample

250 cm³ flask containing *exactly* 90 cm³ sterile distilled water

5 test tubes containing *exactly* 9 cm³ sterile distilled water

10 tubes of sterile, melted, cooled N.A. (15 cm³)

10 sterile petri dishes

Sterile pipettes: 1 x 10 cm³

6 x 1 cm³

Method: Use Fig. 61 as a guide and refer to page 69 Chapter 4 on the method and precautions to observe when making decimal dilutions. Having made the decimal dilutions, using sterile pipette 6 the others having been discarded into a boiling water bath, put 1 cm³ of the 1/10⁶ water dilution into each of two petri dishes, previously well labelled. Using the same pipette now transfer 1 cm³ of the 1/10⁵ dilution into each of the two petri dishes and so on for the 1/10⁴, 1/10³, 1/10². Then pour into each plate one tube of the melted cooled N.A., mix it well with the water sample and allow it to set. When all plates are set incubate one of each pair at 37°C and the other at 22°C (room temperature).

Results: On the whole, colonies which grow on the 37°C plates originate from excremental sources, and those on the 22°C plates are soil and water saphrophytes. Colonies will grow both on the surface and in the agar. Surface colonies will tend to be the larger. Count the colonies in the usual way (See page 65 Chapter 4) and calculate the numbers per 100 cm³ in the original sample.

Using selective media such as MacConkey's medium coliforms only can be counted.

Experiment 41

To identify other organisms in water

On MacConkey agar faecal streptococci produce minute red

colonies. They can also be demonstrated by inoculating blood agar, on which they grow well. The colonies tend to be small and discrete and are often surrounded by a zone of haemolysis. The streptococcal morphology can be identified by staining.

Clostridia, also commensals of the intestine are often in faecally polluted water. They may be isolated by inoculation of ordinary nutrient media or Robertson's meat medium and incubated under anaerobic conditions. In addition if about 50 cm^3 of the water sample is added to sterile litmus milk (100 cm^3) heated to 80°C for 15 minutes to destroy non-sporing organisms, the top sealed with sterile liquid paraffin and incubated for 5 days at 37°C, *Clostridium welchii* may be shown to be present by the formation of a 'stormy clot'. This may develop earlier than five days. The reaction is due to the strong saccharolytic activity of the organism.

Swimming bath water

The tests to estimate the purity of the water are:

    the number of viable bacteria present by plate counts
    the presumptive coliform test
    the differential coliform test

The quality should be high and similar to drinking water.

(Report by the Ministry of Health (1951). *The Purification of the Water of Swimming Baths.* Reference number 6).

10 Examination of milk and food stuffs

Milk

Milk is a natural product, the perfect food for animals, human beings and micro-organisms.

Milk naturally contains quite a number of organisms which are responsible for some of the changes which can take place therein. In healthy cows the milk as secreted is sterile, but the flora is picked up from the teats, from handling of the cows udders and from milking equipment. Most of these micro-organisms are harmless to human beings and their presence is often encouraged in the production of many milk products. However milk can be the means by which disease is spread throughout the community and so milk is subjected to rigorous routine tests to ensure that this is not the case.

Some organisms which occur in milk and which are harmless are: Lactobacillus, *Streptococcus lactis*, some Enterobacteraceae, sporing bacteria, mould spores. All these organisms can contribute to the spoilage of the milk.

Some organisms which can occur in milk and are pathogens are; *Mycobacterium tuberculosis* (causes T.B.), *Brucella abortus* (causes brucellosis), *Streptococcus faecalis* (pyogenic infections). The following series of experiments outline approaches which could well be elaborated and investigated in much greater depth.

Experiment 42
Examination of the flora of milk by staining

A smear of the milk is made in the usual way, (See page 82 Chapter 5) and stain with methylene blue. Clear results can be obtained if the film is defatted after drying by momentarily flooding with xylol followed by immersion in 95% alcohol and then thoroughly washed with water prior to staining. Or, make, de-fat and stain it by Gram's method. This will show the presence of Gram positive cocci (Streptococci) and Gram positive rods, (Lactobacilli). Gram negative rods would not normally be present in milk and would indicate contamination.

Experiment 43
Examination of milk for motile organisms in wet preparation

This is carried out in the normal way, using a diluted sample. The presence of motile organisms would indicate probable pollution of the milk since most of the natural flora of milk are non motile.

- Experiment 44
Examination of the activity of the flora of milk
   (i) Using methylene blue: The normal flora of milk can be demonstrated macroscopically by allowing this flora to reduce methylene blue, the greater the number of organisms present the quicker will be the reduction of the methylene blue.
   Materials: 1 tube boiled milk, 10-15 cm^3
                  1 tube unboiled milk, 10-15 cm^3 (in a sterile tube)
                  Methylene blue (dilute solution)
   Method: Add to each tube about 1 cm^3 methylene blue. Incubate at 37°C preferably in a water bath. Observe at half hour intervals. (A more exact estimation of the number of organisms in a sample of milk is described later—Experiment 46).
   (ii) Using a pH indicator: Since the flora of milk in their metabolism produce acid, their activity can be macroscopically demonstrated in a similar way to that above using, instead of methylene blue, a pH indicator and watching its progressive colour change.

Experiment 45. Culture of the flora of milk
   Ordinary N.A. inoculated with milk will support the flora of milk. Tomato peptone agar adjusted to pH 5·0 and milk agar are good media on which to culture Lactobacilli.
   Milk is treated in different ways to achieve different ends:
   Sterilised milk: This has been treated up to or above 212°F for a period of not longer than 30 minutes, after which there should be no vegetative organisms present although, since some spores are very resistant, there may be some spores present.
   Pasteurised milk is treated in a way such that delicate pathogenic organisms, if present, are killed, but that the essential nature of the milk is unchanged. Pasteurisation is done in two ways:
the flash method: 161°F / 15 seconds,
                  (H.T.S.T. = high temperature, short time)
or   145°F / 30 minutes.
                  (L.T.H. = low temperature, holding.)
   Untreated milk is taken from tuberculosis free herds that is, cattle which have passed the tuberculin test. This milk may contain high numbers of organisms, and spoils fairly rapidly.
   U.H.T. (Ultra heat treated): Heat treated at 133°C for 1 second, and immediately put in sterile containers which are then sealed. This milk is sterile.

Experiment 46
To estimate the number of viable bacteria in a sample of milk
   Materials: 1 bottle of milk, to be examined

4 sterile bottles each containing exactly 90 cm³ sterile water

4 sterile 10 cm³ pipettes

1 sterile 1 cm³ pipette

3 empty sterile petri dishes

3 tubes sterile melted, cooled N.A. (approximately 10 cm³)

Method: Make serial decimal dilution of the milk using the flasks of water and the 10 cm³ pipettes. With the 1 cm³ pipette transfer 1 cm³ of the 1/1000 dilution into each of the 3 petri dishes. Add one tube melted cooled N.A. to the first and mix it with the milk by rotating the plate carefully. Repeat the procedure for the other two plates. Allow all plates to set, incubate at 37°C for 2 days.

Results: Count the colonies, calculate the mean per plate, multiply by the dilution and record the result as 'No. viable bacteria per cm³ of milk'. (See Fig. 62)

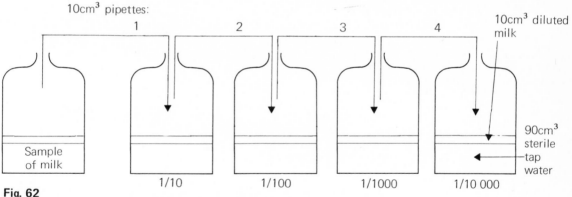

Fig. 62
Estimation of the number of viable bacteria in a sample of milk

Experiment 47

To see whether the milk contains coliform organisms (indicative of faecal contamination)

Materials: The four bottles of diluted milk from the last experiment,

1 x 1 cm³ sterile pipette

12 test tubes containing single strength MacConkey broth with Durham's tubes.

Method: Transfer 1 cm³ of the 1/10 000 dilution into the first three tubes of MacConkey broth. Using the same 1 cm³ pipette transfer 1 cm³ of the 1/1000 dilution into each of the next three broths, then 1 cm³ of the 1/100 into each of the next three broths, and finally 1 cm³ of the 1/10 dilution into each of the last three MacConkey broths. Incubate all 12 tubes at 37°C for 48 hours.

Results: Examine the tubes for acid *and* gas production. Find out the smallest amount of milk which yields acid and gas. Obviously the smaller the amount of milk which yields acid and gas the greater is the degree of contamination. Positive tubes are the result of the growth of the progeny of at least one initial coliform, and so the dilution factor gives an indication of the minimum number of coliforms per cm^3.

In the routine examination of milk, tests which are quicker to perform and which yield quicker results and give an indication of the microbiological status of the milk are favoured. These tests are:

the methylene blue reduction test

the phosphatase test

the turbidity test

Experiment 48
To carry out the methylene blue reduction test on milk

The micro-organisms which are present in milk will reduce M.B. and turn it colourless, the greater the number of organisms, the quicker will be the reduction. This test can be done using all types of milk, except sterilised milk.

Materials: Methylene blue 1/300 000 dilution

3 sterile test tubes and their stoppers

2 sterile 1 cm^3 pipettes

1 sterile 10 cm^3 pipettes

Milk sample

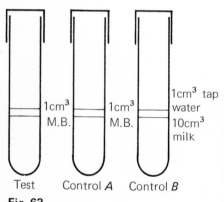

Test   Control *A*   Control *B*

Fig. 63
M.B. test on milk

Method: Set up the test and the controls as shown in Fig. 63. To each tube add exactly 10 cm^3 milk. Place the second and third tubes in a boiling water bath for 3 minutes to destroy the natural reducing system of the milk. To test add 1 cm^3 M.B., to the second tube, control *A* add 1 cm^3 methylene blue, to the third tube, control *B*, add 1 cm^3 tap water. Mix all the tubes well and place them in a 37°C water bath and note the time. Examine each tube every half hour.

Results: The test will gradually turn white and can be compared with the two controls.

Control *A*. This remains blue and the test can be compared with it to see when reduction starts.

Control *B*. This is colourless: comparison of the test with this indicates when the test is complete.

A reduction time of over 4½ hours indicates good milk, one of less than 2½ hours indicates poor milk which contains a large number of bacteria.

M.B. test at 37°C—Raw milk

Reduction time in hours at 37°C	Plate count — numbers/cm³ milk
9	1 400
7	15 500
5	178 000
3	1 995 000
1	22 400 000

'(From *The Milk (Special Designations) Regulations 1963,* S.I. No. 1571. By permission of HMSO.)'

Experiment 49
To carry out a turbidity test on sterilised milk

Materials: 1 sterile 50 cm³ flask
1 sterile 20 cm³ pipette
1 sterile 5 cm³ pipette
4 grams ammonium sulphate
Test milk
Boiling water bath
Test tube
Funnel and filter paper

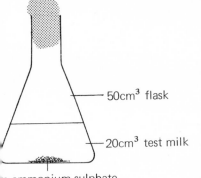

— 50cm³ flask

— 20cm³ test milk

ammonium sulphate.

g. 64
rbidity test on milk

Method: Place 20 cm³ of milk in the flask, add the 4 gm ammonium sulphate, and shake the flask for 3 minutes. Stand the flask on the bench for 5 minutes and then filter it. Take 5 cm³ of the filtrate, put it into a test tube and stand it in the boiling water bath for 5 minutes, cool it and examine it for turbidity (See Fig. 64).

Results: Sterilisation of milk denatures the soluble proteins of milk so that they do not precipitate with ammonium sulphate. The absence of turbidity therefore indicates an adequately sterilised milk.

Ice-Cream
The examination of ice-cream can be carried out in the same way as milk is examined.

Yoghurt
Experiment 50
To make yoghurt

Use homogenised milk and heat to 90°C and hold for a minute or two *or* to 70°C for 30 minutes. Cool it to 40°–42°C and inoculate the milk with 0·5% by volume, of a freshly grown culture of *Lactobacillus bulgaricus* and *Streptococcus thermophilus* grown together. Immediately fill sterile bottles and incubate at 40°C for 2·5 hours until clotting takes place. When firm cool to 5-10°C and hold at this temperature until consumed. The rigidity may be increased by adding gelatin.

Starter culture for yoghurt can be obtained commercially. Stained films of yoghurt (see method for milk page 129) show lactic strepococci and lactobacilli—both Gram positive.

Stock cultures of lactic acid bacteria: Keep in Robertson's cooked meat medium. Sub-culture once a month. Keep cultures a 10-15°C. Initial incubation temperature should be as near to optimum as possible:

Streptococcus cremoris, lactis, faecalis	30-35°C
Streptococcus thermophilus	45°C
Lactobacillus bulgaricus, L. jughort, }	45-50°C
L.lactis }	

Tryptone soy agar, tomato juice agar and broth are also good media for the growth of lactic organisms.

Cheese

Lactic streptococci are used now as 'starters' to allow development of acid in milk to permit the formation of a workable rennet clot.

Examination of Other Foods

Generally speaking other foods can be subjected to an examination similar to that of milk. Virtually all the food that we eat contains a few micro-organisms, it is the type and activity of the organisms present that determines whether the food is edible or not. Among the organisms which contaminate food are moulds, yeasts and bacteria. Some of these organisms may be pathogenic, many are not but cause changes in the appearance and smell of the food which render it unpalatable. Sliminess on the surface of the food may, for example, be due to the luxuriant growth of a capsule forming organism.

Sources of food contamination are varied, the flora originating from animals, sewage, the soil in which the food was grown; from the air, during handling, particularly if fruit and vegetables are damaged the broken surface will allow the prolific growth of food spoilage organisms; from processing in the use of equipment which is not adequately sterilised, in the use of water which is not sterile, failure to reach adequate cooking temperatures throughout the food, in the handling of meat carcases resulting in contamination of the meat from the outside of the animal, and in many other ways. Sometimes the growth of micro-organisms is encouraged in foods. This is because they produce desirable changes in the food, such as by fermentation in wines and beers and other fermented foods and in the production of desirable flavours as in the many cheeses.

Experiment 51
Examination of foods by staining

Wet foods: Such as soups, broths, beer, custards, blanc-manges, sweet fruit juices. Such food, if it is obviously 'off', will on staining show numbers of micro-organisms living in the food. It could be important to de-fat the smear just before staining (Refer to 'Milk' page 129). Spot examination of foods which are not 'off' and which do not normally contain a high indigenous bacterial count will be unlikely to contain enough organisms to be easily visible by staining.

Solid and drier foods: Emulsify a small portion of the test food in sterile saline, or in sterile 0·25 strength Ringer's solution. Use the supernatant fluid to make the smear, fix, and if necessary de-fat the film before staining. Again this method is unlikely to be very successful unless the test food is obviously 'off' or has a natural indigenous flora.

Experiment 52
Examination of foods for motile organisms

Emulsify a small sample of food in sterile saline—try to remove lumps. Examine the fluid as indicated on page 79.

Experiment 53
Culture of the organisms in or on food

Wet foods: These can be used exactly as a broth culture of an organism would be used. A sample can be inoculated into N.B. followed by sub-culture on N.A., isolated colonies obtained and examined, leading to pure culture, identification of the species and perhaps stock cultures. Counts of the numbers of organisms present could also be done.

If fresh sterile foods are obtained, individual species can be inoculated into the foods and changes in the food due to that organism can be observed.

Solid and drier foods: There are several methods for the examination of the flora of solid foods.

(i) Emulsify a small portion of the test food in 5 cm³ sterile saline. Use the supernatant fluid as you would use a broth culture and inoculate N.B. and N.A.

(ii) Sterile swabs moistened in sterile saline can be used to sample the surface flora of solid foods such as fruits and vegetables. The swab can be used directly to inoculate a broth, (in which it is immersed) or the surface of an N.A. plate over which it is rubbed. Counts of the numbers of organisms contaminating the surface of a piece of food can be done (See page 74).

(iii) Agar sausages. These can be used to take impressions of the flora of certain types of food: for example, the outside surface of a piece of meat and the freshly cut internal surfaces (which should be sterile). Similarly with cheese, bread, fish and so on (see page 75).

Experiments which demonstrate the various methods for the preservation of food

(1) Cooking

Cooking kills the contaminating organisms in and on food by wet heat sterilisation. It is easy to demonstrate this effect.

Experiment 54

The near sterilising effects of cooking

Materials: 1 piece of meat
1 sterile scalpel
1 pair sterile forceps
1 sterile empty petri dish
2 N.A. plates
2 test tubes of water, 5-10 cm^3 each
2 N.B. sterile

Method: With the sterile scalpel cut four small pieces of meat (numbers 1,2,3,4) from the surface of the piece of meat. Transfer each to the sterile petri dish. Hold the piece of meat number 1 in the sterile forceps and smear it over the surface of the N.A. plate 1. Label this plate 'Uncooked'. Drop piece of meat number 2 into a tube of water and boil it for 20 minutes and allow it to cool. Extract the piece of meat from the tube, allow the excess water to drip off and then smear it over plate 2. With sterile forceps drop piece of meat number 3 into N.B. label it 'Uncooked'. Drop piece of meat number 4 into water. Boil it for 20 minutes, cool it and then transfer it to N.B. tube 2 label that tube 'Cooked'. Incubate all plates and broths at 37°C.

Results: After incubation there should be visible and demonstrable growth in the media labelled 'Uncooked' but not in the media labelled 'Cooked'. *Note:* Growth on plates after boiling food: (1) 20 minutes boiling will not necessarily kill all bacterial spores, and it is possible to get growth after boiling. (2) Contamination?

(2) Refrigeration

Refrigeration of foodstuffs does not kill their flora but merely slows the growth, so that when food is removed from refrigeration the flora will begin to grow again. In addition to this there are organisms—psychrophils (cold loving) which thrive at low temperatures. These organisms, non pathogens, can bring about changes in appearance and smell which alter its palatability and make it less attractive to eat. An easy way to demonstrate the slowing up of multiplication rate of micro-organisms due to refrigeration is to use milk.

Experiment 55
To examine the effect of refrigeration on the flora of milk
    Materials: Equipment for plate counts as Experiment 46.

    Method: Carry out a plate count on skimmed milk as described in Experiment 46 in this chapter. Then put 10 cm^3 of the milk into each of 4 sterile test tubes. Place 2 in the refrigerator at about 4°C and leave the other 2 at room temperature. Do a plate count on one tube from the refrigerator and one from the room, after 24, and 48 hours. After that bring the refrigerated tubes into room and incubate at room temperature, then perform a plate count.

    Results: All counts compared with the first will show an increase in numbers, but those of the refrigerated milk less so than those at room temperature. When the refrigerated milks have been incubated at room temperature they will show a great increase in numbers showing that the organisms have not been killed or damaged by refrigeration.

(3) Dried food
    Food can be preserved by dehydration since micro-organisms are unable to multiply. Once water is added however the micro-organisms present will be able to multiply.

Experiment 56
To demonstrate the presence of micro-organisms in dried food
    Materials: Dried milk, or other dried food
                    2 sterile N.B.
                    Broths, pipettes, N.A. for a viable count.

    Method: Take samples from a partially used tin of dried milk (or other dried food) Emulsify 1 gram of the sample in sterile broth (9 cm^3), make serial decimal dilutions and perform a count on the numbers present in the dried food. Emulsify 1 gram in sterile broth (9 cm^3) and incubate at 37°C for 24 hours. Perform a viable count on this. This will inevitably show an increase in the number of organisms present, illustrating why partially cooked foods like custards and blanc-manges should not be left overnight out of refrigeration (as they often are).

(4) Polythene packaging
    This method is in common use these days as a protection for perishable foods sold in supermarkets. The polythene does not preserve the food but rather protects it from handling.

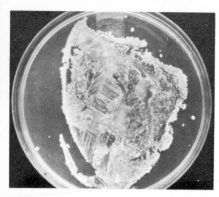

Bacterial growth on a N.A. plate after an impression had been made on it using a piece of vacuum packed bacon. Note the very dense growth after 24 hours incubation at 37°C (about ¼ life size).

Experiment 57

Investigating the polythene packaging around bacon rashers

    Materials: 1 large N.A. plate
                           Bacon rashers

    Method: Bacon rashers are often sold 'vacuum packed'. An interesting comparison can be made between the numbers and distribution of the flora of a freshly cut bacon rasher and one taken from a vacuum pack.

    Pour a large plate (i.e. use the lid of a 7lb biscuit tin and a piece of glass as the top). (See page 35). Impress each bacon rasher onto the surface of the N.A. Remove each, label the plate, and incubate at 37°C for 24 hours.

    Results: It is possible that the vacuum packed rasher will have a greater flora than the other rasher. The great value of these packages is not, therefore, that they keep the bacon nearly sterile but that they protect the food from contamination by food poisoning organisms when handled by customers.

Experiment 58

(5) Use of osmotic pressure

    Most bacteria are sensitive to the high osmotic pressures due to strong salt or sugar solution and are unable to grow in them. This sensitivity may be easily demonstrated in either of the following ways:

    Methods:

    — (i) Using gradient plates.

    Using a tube of N.A. containing 10% salt, make a gradient plate of this with a layer of plain N.A., so that the salt concentration range is 0-10%. Streak the test organisms parallel to the gradient.

    (ii) In Nutrient broth.

    Make a series of nutrient broths containing graded concentrations of salt. An easy way to do this is to make a stock solution of 10% NaCl in nutrient broth, and add this aseptically in appropriate volumes to tubes of sterile N.B. (See Fig. 65). To each of the 11 tubes add a standard inoculum of the test organism and incubate all tubes at 37°C for 24 hours.

Fig. 65
Tubes of varying osmotic pressure

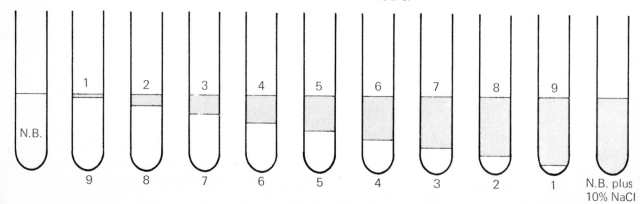

Results: The tolerance of the organisms to salt will be visible: growth will appear in those tubes, or on the gradient plate where the concentration of salt can be tolerated and not beyond.

Caution: One of the food poisoning organisms, *Staph. aureus*, has a high salt tolerance and this is in fact one of the ways in which it is isolated. Salted foods may be spoiled by salt tolerant organisms.

(6) Tinned food

If the process is carried out properly there should be no organisms present inside the tin.

Sampling tinned food: The top of the unopened tin should be thoroughly cleansed and the sterilised, and then opened with a sterile opener. The contents can then be turned into a sterile container and the flora tested in ways suggested earlier. Some organisms can thrive in the anaerobic conditions inside a tin and can be very dangerous species. So it is recommended that unless you are very sure of your technical ability, you do not attempt to culture the contents of an 'off' tin but instead get in touch with the public health laboratories and leave them to decide what to do.

The danger of reheating food

The danger involved is that 'left-overs' will have been exposed to contamination, and will therefore act as the medium for the growth of the micro-organisms, if the food is not immediately refrigerated. These will have increased in number, and reheating alone will not be sufficient to kill them. If it is prolonged it will encourage rapid cell multiplication so that, when eaten, the 'left overs' could lead to food poisoning.

Foods produced as a result of microbial activity.

Ginger beer
Beer, and wines
Fermented foods such as sauerkraut, and silage
Milk products

Summary of some important bacteria

Family Pseudomonadaceae:

Rod like bacteria, usually motile, gram negative may form water soluble pigments (green or brown) or non water soluble pigments (yellow or red). Two genera of importance:

Genus *Pseudomonas:* These are very widely spread organisms which can easily be cultivated in the laboratory on ordinary media. Some species may be responsible for surface changes in colour of food. (The pigment pyocyanin, bluish green, produced by some members of this group may be extracted from broth culture using chloroform).

Genus *Acetobacter:* Some species of this genus are of great value in the manufacture of vinegar. Other members, since they grow well on many food stuffs are responsible for surface changes such as colour and slime.

Both the genus *Pseudomonas* and genus *Acetobacter* grow well at low temperatures and are often responsible for spoilage of food in refrigeration.

Family Enterobacteraceae:

Many species grow in the intestinal tract of man and animals and hence give the name to the family. They are rod shaped, gram negative, often motile, ferment glucose and often other carbohydrates.

Genus *Escherichia* and Genus *Klebsiella:* These are the 'Coliforms' whose presence on or in food indicates contamination by sewage.

Genus *Erwinia:* These are responsible for some soft rots in plants e.g. *Erwinia carotovora* produces soft rots in carrots.

Genus *Serratia:* These produce flavours in foods, and are responsible for 'ropiness' in milk, and infections of bread. *Serratia marcescens* is the most common and produces a bright red pigment.

Genus *Proteus:* These are very motile organisms which are sometimes associated with spoilage of meat and of seafoods.

Genus *Salmonella* and Genus *Shigella:* These are enteric pathogens causing typhoid and dysentary like conditions and food poisoning. They may be transmitted by water and by foods. They will get into foods by unhygenic handling and use of non sterile water. They originate from faeces of man and animals, from infected carcases, eggs and milk.

Family Micrococcaceae

Spherical gram positive cells, many produce non water soluble yellow or orange pigments. Non motile.

Genus *Staphylococcus:* Some of the members of this genus cause food poisoning (*Staph. aureus*) because they produce a poison: an enterotoxin. Foods such as cooked meats, milk products, custards, artificial creams are often contaminated in their preparation and due to their method of sale, incubation in a shop window or counter, the numbers of organisms increase and give rise to food poisoning when eaten.

Genus *Sarcina:* Many species in this genus produce pigmented colonies. *S. lutea:* causative organism of 'yellow spot' on meat carcases.

Family Lactobacillaceae

These are the lactic acid bacteria which need complex foods, and which give fairly poor growth on ordinary media. The family comprises gram positive rods and cocci.

Genus *Streptococcus:* These are gram positive cocci typically arranged in chains. The 'pyogenic group' which may occur in foods and milk cause sore throats of varying severity, including scarlet fever, the 'Viridans group' include *Strep. thermophilus* which is important in cheese making, and the 'Lactic group' whose members are often important in making various milk products e.g. *Strep. lactis, Strep. cremoris.*

Genus *Lactobacillus:* These are gram positive rods, they ferment carbohydrates with the formation of lactic and other acids. They are very important in the microbiology of milk and milk products.

Family Corynebacteria

In appearance these are beaded and banded organisms which occasionally occur in foods. An important member of the family is *Corynebacterium diphtheriae* causative of diphtheria and which may be transmitted in foods.

Family Bacillaceae

These are gram positive sporing organisms which occur widely.

Genus *Bacillus:* These are gram positive aerobic organisms active biochemically, which can curdle milk and which produce flatsours in foods, that is, souring of the food without gas production. Some species can cause ropiness in bread, and many others are associated with the spoilage of food.

Genus *Clostridium:* These are gram positive, strictly anaerobic, sporing organisms. They are active biochemically and can be responsible for undesirable changes in foods, especially canned foods. In addition some strains produce very potent exotoxins which are extremely poisonous. The important members of this group are *Cl. botulinum, Cl. tetani,* and *Cl. welchii.* The latter is associated with food poisoning and originates in the bowel.

Family Actinomycetales

These are the mould-like bacteria having branching colonies and mould like growth. They can cause spoilage producing musty flavours in the food.

11 Examination of soil and air, and the transference of disease

Soil

It is important to realise that the soil contains very many micro-organisms other than the bacteria and the moulds that this chapter aims to demonstrate (see page 74).

Experiment 59
Examination of the flora of soil by staining

Make an emulsion of soil in water or in N.B. and use the supernatant fluid to make the smear. Stain the smear by Gram's method and with a spore stain. These methods will demonstrate the great variety of soil micro-organisms.

Experiment 60
Examination of the motility of soil flora

A suspension of soil in N.B. or water is examined for the presence of motile organisms in the usual ways (See page 79). It will show the large numbers of motile organisms which live in the soil water.

Experiment 61
Culture of the soil flora

Emulsify samples of the soil in sterile saline or in N.B. Use some of this to inoculate N.B. and plates of N.A. and malt agar (or other media suitable for moulds). Incubate these media aerobically and anaerobically at 22°C and at 37°C. This technique will inevitably yield a good crop of organisms which can be isolated and identified showing the variety of organisms in the soil. It would be of great interest to compare the flora at different soil depths, and in different soils. The pour plate method for the isolation of organisms is a good technique to use in the case of a very mixed inoculum such as soil.

Examination of soil for the presence of antibiotic producing organisms
(See page 112).

Experiment 62
Counts of the numbers of organisms present in soil samples

Method: Emulsify 1 gram of soil in 10 cm^3 N.B. and use this as the 10^{-1} dilution. Then proceed as for a normal count, doing it in duplicate and incubating one set at 37°C and the other at 22°C. Use of N.A. at pH 7·5 will exclude the growth of moulds. Since so many herbicides, pesticides, fertilisers and so on are used on soil it would be an interesting project to study their effect on soil flora.

Some of the bacteria found in the soil:

The variety of bacteria found in the soil is considerable. Commonly found are

Genus *Bacillus*
Genus *Clostridium*
Genus *Nitrobacter*
Genus *Pseudomonas*
Genus *Azotobacter*
Genus *Nitrosomonas*
Genus *Streptomyces*
Genus *Actinomyces*
Family Enterobacteraceae

Nitrogen fixation

Some micro-organisms fix atmospheric nitrogen, playing a very important part in the nitrogen cycle. Some of these organisms are non-symbiotic such as members of the Genera *Azotobacter, Clostridia, Aerobacter,* and some are symbiotic for instance Genus *Rhizobium.*

Experiment 63
Examination of root nodules

The Rhizobium family form root nodules in the roots of legumes: peas, beans, legumes, clovers, lupins, alfalfa. Select a root nodule from the root of a legume. Wash it well. Cut off a small portion with a sterile scalpel and crush this on a slide. Make a smear of the portion using a second slide. Fix, and stain with N.B. On examination it will show some of the various forms of Rhizobium, known as bacteroids and have forms similar to letters and are known as, *X, Y* and *V* forms.

Experiment 64
Isolation of Rhizobium from nodules

Select a nodule and wash it well in tap water. Then soak it well in 1 : 100 phenol for 20 minutes, and then rinse it in sterile distilled water. Cut the nodule open using a sterile scalpel and homogenise it and use the homogenate to inoculate a plate of medium: trypticase soy agar, or potato dextrose agar, or Czapek Dox medium, or N.A. made with yeast extract and adjusted to pH 7·5 or mycological agar. Incubate at 20°C. Rhizobium species is gram negative, aerobic, non sporing and usually motile, pleo-morphic in form showing great variety in cell shape varying from to cocci to rods.

Experiment 65
Inducing nodule formation in legumes

Take the seeds of a legume, such as clover, and sterilise them in disinfectant. Make up a quantity of Crone's nitrogen free medium:

KCl	5 gm
K_2HPO_4	1·25 gm
$CaSO_4.2H_2O$	1·25 gm
$MgSO_4.7H_2O$	1·25 gm
$Ca_3(PO_4)_2$	1·25 gm
$Fe_3(PO_4)_2$	0·01 gm

by grinding up the salts well together. Add 1·5 grams of this to 1 litre of water and add 8 grams of agar. Autoclave this to sterilise it. Dip the sterile legume seeds into a culture of Rhizobium and drop the seeds into tubes of Crones medium and allow them to grow. Nodules will form on the roots. Examine the nodule after about 5 weeks.

Experiment 66
Growth of a non-symbiotic nitrogen fixing organism—*Azotobacter*

These non-symbiotic bacteria can be isolated from soil by the inoculation of a medium such as Crone's, solidified with 1·5% agar and containing a trace of molybdenum which is essential to these organisms, and adjusted to pH 7·2. The growth of *Azotobacter* is inhibited by the presence of organic compounds in the medium. *Azotobacter* motile pleomorphic organisms can be isolated from a number of soil types particularly well aerated arable ones. They will grow well in any nitrogen free salt solution at neutral or slightly alkaline pH.

Examination of Air

The air contains micro-organisms in suspension. These organisms are not actively living and dividing but rather are existing in their vegetative form, or as spores, and will be able to grow and multiply when they are deposited in a suitable medium. They can exist in this state of 'suspended animation' for various lengths of time, some species being able to exist for many years. All air contains some micro-organisms their numbers varying from place to place. Near to the ground the numbers are high, further away from the ground the numbers decrease but micro-organisms have been isolated from samples of the atmosphere taken from extremely high altitudes. The organisms are derived from human and animal bodies, from dust and soil and many other places. Often organisms which have laid dormant for long periods in dust before being suspended in air are responsible for outbreaks of infection. So it is convenient not to make too definite a separation of the study of air borne organisms from the study of the transference of disease.

Experiment 67
Sampling of air

Expose plates of nutrient medium suitable for the growth of bacteria and moulds to the air; or blow onto the medium by means of a bicycle pump. The plates, or broths after incubation will show growth not unlike that obtained from samples of dust or from surfaces, human, animal and inanimate objects. Results can be plotted as block graphs comparing place with place (e.g. W.C., canteen, open country and so on)

Experiment 68
A second method for sampling the suspended flora of air

Air can be pumped through a sterile sand or cotton wool filter which is later washed in nutrient broth and from which a culture or a plate count can be made. It is possible that this technique might be used in conjunction with testing air which has been exposed to ultra-violet light or which has been exposed to aerosol disinfectant sprays.

Transmission of disease

The methods by which disease causing organisms are transferred from one place to another are broadly speaking by means of inanimate objects and through the agency of living organisms. *Fomites* are objects which become contaminated from an infected source and which are liable to be touched or used by people or animals who will then become infected. Examples of fomites are door knobs frequently handled; pens and pencils sucked, licked and chewed, and borrowed; plates, knives, forks and spoons, inadequately washed and used for meals; handkerchiefs; toys; bed-linen of the ill; and indeed any object which in its use passes from one person to another or from one animal to another. Food is an obvious source for the transmission of disease, taken from one source, prepared possibly in several stages from the factory to the house and liable to infection at all those stages, and eaten by others. Hands are often instrumental in the transmission of disease: hands not washed after use of the toilet, fingers sucked and nails bitten, dirty hands used to prepare food, hands which handle money. *Mucus* and droplets of *saliva* become suspended in the air due to coughs and sneezes and are inhaled by unsuspecting passers-by, or eventually come to rest in dust to be a source of infection at a later time. Dust not only preserves organisms derived from saliva and mucus but also organisms derived from faeces and from soil. Direct physical contact between humans passes on infection of both mild and of more severe natures, as does direct contact between man and animals, and between animals. In addition there is a state of infection in which a person can be responsible for the spread of a species of organism but not suffer from its pathogenic effects themselves. This state is known as the 'carrier state', a person acting as a reservoir of infection. Arthropods also are responsible for transmission of disease-causing-organisms in several ways. They carry bacteria on their feet and on their bodies from one place on which they have crawled to another. Their bodies often harbour organisms which are spread by means of their faeces dropping onto foods. Biting insects transmit organisms in their saliva and from their mouth-parts. Some pathogenic organisms are spread from animal bites and some pathogens can be spread through blood transfusion, though the latter is very rare.

Experiments on the transference of disease

Knowledge of the methods of the transference of disease enable the teacher to set up experiments to demonstrate this. It is a good idea to use a non-pathogenic organism as an indicator such as *Serratia marcescens,* an organism which produces bright red colonies and is easily recognisable. Broadly speaking the method to adopt is to inoculate a test surface, a fomite, with the indicator organism, to allow normal handling of that fomite and to follow the spread of the organisms by test sampling of surfaces to see where they have got to. Use of agar sausages would be of great help in this. Demonstration of the spread of an organism by means of an insect vector can be performed in the following way. Introduce an insect onto a plate culture of an indicator organism and subsequently remove and introduce it to a second plate of medium and allow it to walk over that medium. Incubation of the plate should show the tracks of the creature. Alternatively introduce a freshly captured blowfly to a plate of N.A. and allow it to crawl about, remove it after a short while and then incubate the plate.

'Coughs and sneezes spread diseases' is quite true, and coughing and sneezing over plates of medium will soon show how the organisms are disseminated.

Experiment 69

Koch's postulates are a series of rules which allow the identification of an organism as being the causative agent of a disease. These postulates state that (1) the species of organism should always be found in a particular disease, that (2) it be isolated in pure culture and (3) it should be capable of producing the same disease in a susceptible animal (or plant). An easy way to demonstrate Koch's postulates is to allow a carrot to go slimy and from the slime to isolate the organisms in pure culture on solid media. When this has been done pieces of sterile carrot (Refer back to the chapter on sterilisation) should be inoculated with the organism isolated from the slimy carrot. For Koch's postulates to be proven the same symptoms must be present.

Experiment 70
Demonstration of the effect of disease on a population using the model of bacterium and bacteriophage

It is a commonly observed fact that when a disease attacks a population no matter how virulent that disease is not all the population suffer from it, and some survive. This is due to various factors one of which is the natural resistance of some individuals to the infective agent. A *model* of how some members of a population are resistant to an infective agent can be set up using bacterial virus-bacteriophage to represent the pathogen, and a susceptible host bacterium as the population to be infected.

Theory: Bacteriophage, known as 'phage, is the name given to very minute virus particles which can invade susceptible bacterial cells, multiply within them exhaust the cell contents and cause the lysis, and hence the death, of the bacteria. If drops of phage culture are dropped onto a lawn of a host bacterium clear zones of near- or no-growth will show after incubation. These zones, or plaques, represent widespread cell death and are due to invasion by phage, phage multiplication and cell lysis.

Some bacteria can be invaded by the 'phage but not lysed by it. These bacteria can give rise to a clone of 'lysogenic' cells, that is cells containing the 'phage in a latent state, which may at a later date release virulent 'phage particles.

Demonstration of the presence of 'phage: 'Phage are often present in natural materials such as soil, water, mud and sewage. The 'phage types on which most work has been done are those affecting the organism *Escherichia coli*, but many other organisms have been shown to be susceptible to their own specific 'phages.

Materials: 250 cm^3 N.B. in a 3 litre flask
Sample from which 'phage is to be isolated:
10-20 gm faeces *or*
10-20 gm soil *or*
100 cm^3 river water and so on.

Young broth culture of the organisms for which 'phage is required

e.g. *Escherichia,*
Salmonella,
Pseudomonas

Sterile 1 cm³ pipette
Filter, and filter paper (sterile)
N.A. plate
Large loop

Method: Add the sample material to the N.B. and mix it well together. Add 1 cm³ of the culture of organisms for which the phage is required to enrich the phage. Incubate the broth culture overnight in aerobic conditions at 37°C. The next day filter it well through several layers of filter paper (sterile) and keep the filtrate. To test for the presence of phage: On the N.A. plate make several broad parallel strokes of the test organism. Allow the inoculum to sink into the agar. Then with a large loop spot inoculate *onto* the inocula drops of the 'phage filtrate. (Being a virus the 'phage particle will pass through the filter). Incubate overnight.

Results: If 'phage specific to the test organism is present then plaques of lysis will appear on the streaks of culture. Single colonies of 'phage resistant mutants may appear within the plaques.

If this experiment is being used as a model of the spread of an epidemic it is important to remember that the reasons for the resistance of some bacteria to the 'phage are not the same as the reasons for the resistance of a human being to a disease, and that this merely demonstrates the existence of cells naturally resistant to a pathogen.

12 Growth

The development of a culture results in an increase in the bacterial numbers and in the total amount of protoplasm present. Bacteria transferred to a new and favourable medium grow and multiply and soon reach a maximal population. The medium supports this for a time but soon the numbers of organisms present decline and fall off. The factors which limit the growth of organisms in liquid culture are the availability of oxygen (if required), the accumulation of the toxic end products of metabolism and the availability of food.

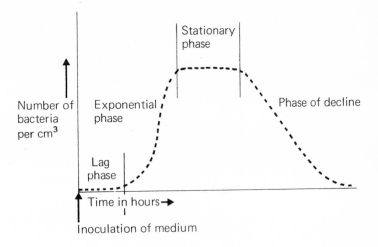

Figure 66 shows the pattern of growth in a liquid medium in a closed environment. There is first of all a *lag phase* during which time the organisms adjust themselves to their new environment; when they have adjusted they grow well and multiply rapidly entering the phase known as the *exponential phase*; as the numbers increase so the quantities of toxic end products of metabolism increase and the food supply decreases. These combined end the exponential phase and the rate of increase is slowed down so that the overall number of organisms remains constant for a while, the *stationary* phase, until the medium is no longer very conducive to growth and the culture enters the phase of decline so that eventually the culture dies. The times for these phases depend on the strain of organism and the cultural environment.

These growth phases can be investigated experimentally but with elementary equipment can be rather laborious, since counts

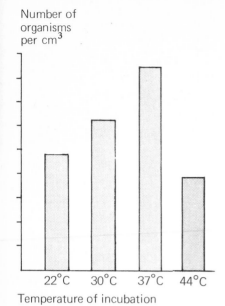

Number of organisms per cm^3

22°C   30°C   37°C   44°C

Temperature of incubation

Fig. 67
The effect of temperature on growth

of the numbers of organisms present at various stages are required.

Experiment 71
To obtain a growth curve for an organism

Inoculate a flask of medium with the test organism e.g. *E. coli* and incubate at 37°C. Mix well and withdraw a sample every four hours to make a plate count and thus eventually obtain readings for a growth curve.

Experiment 72
To investigate the effect of temperature on growth

Inoculate 0·1 cm^3 of broth culture of the test organism into flasks of broth and incubate the flasks at different temperatures: 22°C, 30°C, 37°C, 44°C, and 60°C. Incubate for an equal length of time, withdraw samples and make plate counts on each. Plot the results as shown in Fig. 67.

Experiment 73
To investigate the effect of pH on growth

Inoculate a standard inoculum of the test organism into flasks of broth whose pH varies: 6·0, 7·0, 8·0, for example, and incubate at 37°C. After incubation make plate counts to find the numbers of organisms present.

Experiment 74
To find out the effect of the state of aeration on the growth of a culture

Inoculate 0·1 cm^3 of broth culture of the test organism into each of three flasks of N.B. pH 7·0. Incubate them under the following conditions at 37°C:

well shaken (i.e. aerated).

static

anaerobically (Put the flask of medium in a boiling water bath for ten minutes, cool it and inoculate it and then cover the top with a layer of sterile melted vaseline.) After about 4-5 hours incubation put up plate counts as before.

Experiment 75
To show the effect of nutrient concentration on the growth of a bacterium

Inoculate the test organism into N.B. in flasks which has been especially prepared: one flask with the strength of the nutrients 0·1 of its normal strength, a second with normal strength nutrients, and a third with the nutrients 2, 3 or even 10 times as concentrated as the normal medium. In all media adjust the pH to 7·0 and incubate at 37°C for 24 hours. Then set up plate counts.

Experiment 76
To investigate the effect of high osmotic pressure on an organism

An experiment very similar to Experiment 75 can be set up in which the osmotic pressure of the normal N.B. is adjusted by adding varying quantities of NaCl to the medium, inoculating, incubating, and setting up plate counts.

Appendix

Antibiotics, some suppliers

Beecham Group Ltd.,
Beecham House,
Great West Road,
Brentford, Middlesex.

Burroughs Wellcome,
Wellcome Chemical Works,
Dartford, Kent.

Glaxo Laboratories Ltd.,
Greenford Road,
Greenford,
Middlesex.

British Drug Houses Ltd.,
Poole,
Dorset.

C.I.B.A. Laboratories Ltd.,
Horsham, Sussex.

Autoclave, temperature, pressure, time relationship

Temperature	Pressure in lb/sq.in above atmospheric	Exposure time
100°C	0	90 minutes to kill spore
105°C	2·8	
109°C	5·4	30 minutes
115°C	9·8	
121°C	15·0	15-45 minutes
125°C	19·0	
130°C	24·5	
135°C	30·7	3 minutes only in a pre-vacuum autoclave.

Cultures

A list of the collections of micro-organisms in Great Britain and from whom cultures may be obtained on request.

Bacteria, of medical and veterinary interest:

The Curator, National Collection of Type Cultures,
Central Public Health Laboratory
Colindale Avenue,
London, N.W.9.

A catalogue of organisms is available on request. Only non-pathogenic organisms are issued.

Fungi, plant pathogens, industrial fungi, syllabus 'type' used in education, and isolates of taxonomic significance.

The Assistant Director,
Commonwealth Mycological Institute Collection of Fungus Cultures,
Kerry Lane, Kew, Surrey.

Fungi and Yeasts pathogenic to animals.

The Director,
Mycological Reference Library,
London School of Hygiene and Tropical Medicine,
Keppel Street,
Gower Street,
London, W.C.1.

Wood rotting fungi.

The Director,
Forest Products Research Laboratory,
Princes Risborough,
Aylesbury,
Bucks.

Yeasts, non-pathogenic.

The Director,
National Collection of Yeast Cultures,
The Brewing Industry Research Foundation,
Nutfield,
Redhill, Surrey.

Industrial bacteria, bacteria for microbiological assay, etc.

The Curator,
National Collection of Industrial Bacteria,
Department of Scientific and Industrial Research,
The Torry Research Station,
135 Abbey Road,
Aberdeen.

Bacteria, pathogenic to plants.

The Curator,
National Collection of Plant Pathogenic Bacteria,
Ministry of Agriculture, Fisheries and Food,
Plant Pathology Laboratory,
Milton Road,
Harpenden, Herts.

Bacteria of Milk and Milk products.

The Director,
National Collection of Dairy Organisms,
National Institute of Research in Dairying,
Shinfield,
Reading, Berks.

Bacteria, marine.

The Director,
National Collection of Marine Bacteria,
Department of Scientific and Industrial Research,
The Torry Research Station,
135 Abbey Road,
Aberdeen.

Algae and protozoa.

The Curator,
Culture Collection of Algae and Protozoa,
Botany School,
The University,
Downing Street,
Cambridge.

Cultures

Suggested list of organisms to have as stock in the laboratory.

Staphylococcus albus	Gram positive coccus, bunches
Staphylococcus aureus*	Gram positive coccus, bunches
Streptococcus viridans	Gram positive coccus, chains
Neisseria catarrhalis	Gram negative coccus, pairs
Corynebacterium xerosis	Diphtheroid bacillus
Mycobacterium phlei	Commensal.
Bacillus subtilis	Gram positive, saphrophytic, sporer, aerobic
Clostridium welchii*	Gram positive, saccharolytic, sporer, anaerobic
Clostridium sporogenes	Gram positive, proteolytic, sporer, anaerobic
Salmonella type*	Gram negative, Rod
Shigella type*	Gram negative. Rod
Escherichia coli	Gram negative. Motile
Proteus vulgaris	Gram negative. Motile
Pseudomonas fluorescens	Gram negative. Produces a fluorescent pigment
Lactobacillus acidophilus	Gram positive rod. Sours milk
Azotobacter	Fixes N_2 in the soil
Rhizobium	Fixes N_2 symbiotically
Serratia marcescens	Gram neg. rod. Bright red colony
Sarcina lutea.	Gram pos. Cocci in packets of 8. Yellow colony

*—extra care with these—they should be treated as pathogens.

Dyes

Concentrations to use in experiments

Brilliant green and other organic dyes inhibit in very low concentrations

   e.g. $1/10^7$

Crystal violet. Use the range 1/100 000, 1/200 000, 1/400 000
   1/800 000, 1/1 000 000
   1/500 000 is inhibitory to most Staphylococci.

Proflavine and acriflavine use 1/1000 up to 1/100 000.

Film loop

Bacteriological techniques
   Part 1   81-004
   Part 2   81-005

The Ealing Corporation,
Cambridge,
Massachusetts. Marketed in this country.

Probability tables for the estimation of the numbers of coliform organisms in milk

These indicate the estimated number of bacteria of the coliform type present in 100 cm^3 of water. (Computed by McGrady 1918, amended by Swaroop (1938)).

'(From Reports on Public Health and Medical Subjects No. 71, *The Bacteriological Examination of Water Supplies.* Reproduced by permission of HMSO.)'

Quantity of water put in each tube:	50cm³	10cm³	
Number of tubes used:	1	5	
Number of tubes giving a positive reaction			*Most probable number of coliforms in 100 cm³ of original water*
	0	0	0
	0	1	1
	2	2	2
	0	3	4
	0	4	5
	0	5	7
	1	0	2
	1	1	1
	1	2	6
	1	3	9
	1	4	16
	1	5	18+

Quantity of water put in each tube	10cm³	1cm³	0·1cm³	Most probable number of coliforms in 100 cm³ of original water
Number of tubes used	1	5	5	
	0	0	0	0
	0	0	1	1
	0	0	2	2
	0	1	0	1
	0	1	1	2
	0	1	2	3
	0	2	0	2
	0	2	1	3
	0	2	2	4
	0	3	0	3
	0	3	1	5
	0	4	0	5
	1	0	0	1
	1	0	1	3
	1	0	2	4
	1	0	3	6
	1	1	0	3
	1	1	1	5
	1	1	2	7
	1	1	3	9
	1	2	0	5
	1	2	1	7
	1	2	2	10
	1	2	3	12
	1	3	0	8
	1	3	1	11
	1	3	2	14
	1	3	3	18
	1	3	4	20
	1	4	0	13
	1	4	1	17
	1	4	2	20
	1	3	3	18
	1	3	4	20
	1	4	0	13
	1	4	1	17
	1	4	2	20
	1	4	3	30

Number of tubes giving positive reaction

10cm³	1cm³	0.1cm³	MPN
1	4	4	35
1	4	5	40
1	5	0	25
1	5	1	35
1	5	2	50
1	5	3	90
1	5	4	160
1	5	5	180+

Quantity of water put in each tube	$10cm^3$	$1cm^3$	$0.1cm^3$	
Number of tubes used	5	5	5	
Number of tubes giving positive reaction				*Most probable number of coliforms in 100 cm³ of original water*
	0	0	0	0
	0	0	1	2
	0	0	2	4
	0	1	0	2
	0	1	1	4
	0	1	2	6
	0	2	0	4
	0	2	1	6
	0	3	0	6
	1	0	0	2
	1	0	1	4
	1	0	2	6
	1	0	3	8
	1	1	0	4
	1	1	1	6
	1	1	2	8
	1	2	0	6
	1	2	1	8
	1	2	2	10
	1	3	0	8
	1	3	1	10
	1	4	0	11
	2	0	0	5
	2	0	1	7
	2	0	2	9
	2	0	3	12
	2	1	0	7
	2	1	1	9
	2	1	2	12
	2	2	0	9

Quantity of water put in each tube	$10cm^3$	$1cm^3$	$0.1cm^3$	Most probable number of coliforms in $100cm^3$ of original water
Number of tubes used	1	5	5	
	2	2	1	12
	2	2	2	14
	2	3	0	12
	2	3	1	14
	2	4	0	15
	3	0	0	8
	3	0	1	11
	3	0	2	13
	3	1	0	11
	3	1	1	14
	3	1	2	17
	3	1	3	20
	3	2	0	14
	3	2	1	17
	3	2	2	20
	3	3	0	17
	3	3	1	20
	3	4	0	20
	3	4	1	25
	3	5	0	25
	4	0	0	13
	4	0	1	17
	4	0	2	20
	4	0	3	25
	4	1	0	17
	4	1	1	20
	4	1	2	25
	4	2	0	20
	4	2	1	25
	4	2	2	30
	4	3	0	25
	4	3	1	35
	4	3	2	40
	4	4	0	35
	4	4	1	40
	4	4	2	45
	4	5	0	40
	4	5	1	50
	4	5	2	55
	5	0	0	25

Number of tubes giving positive reaction

Number of tubes giving positive reaction			Most probable number of coliforms in 100cm³ of original water
5	0	1	30
5	0	2	45
5	0	3	60

Quantity of water put in each tube	10cm³	1cm³	0·1cm³
Number of tubes used	1	5	5

Number of tubes giving positive reaction			Most probable number of coliforms in 100cm³ of original water
5	0	4	75
5	1	0	35
5	1	1	45
5	1	2	65
5	1	3	85
5	1	4	115
5	2	0	50
5	2	1	70
5	2	2	95
5	2	3	120
5	2	4	150
5	2	5	175
5	3	0	80
5	3	1	110
5	3	2	140
5	3	3	175
5	3	4	200
5	3	5	250
5	4	0	130
5	4	1	170
5	4	2	225
5	4	3	275
5	4	4	350
5	4	5	425
5	5	0	250
4	5	1	350
5	5	2	550
5	5	3	900
5	5	4	1600
57	5	5	1800+

[Reproduced by permission of H.M.S.O.]

0–20 correct to the nearest unit
20–200 correct to the nearest 5
200–250 correct to the nearest 25
Above 500 correct to the nearest 50

Ringer's solution

NaCl	9·0gm
$CaCl_2$	0·48gm
KCl	0·42gm
$NaHCO_3$	0·2gm

Distilled water to 1000 cm^3

References

Chapter 2
1. Mackie and McCartney's *Handbook of bacteriology.* Edited by Robert Cruikshank. E. and S. Livingstone Ltd. 10th edition.
2. *Oxoid Manual.* 3rd Edition. 1965. Oxoid Ltd.
3. Leifson, E., Cosenza, B.J., Murchelano, R., Cleverdon, R.C. *Motile marine bacteria : I. Techniques, ecology and general characteristics.* J. Bact. **87**, 652, (1964).

Chapter 4
4. Knetman,. A., *A method for the cultivation of anaerobic spore bearing bacteria.* J. Appl. Bact., **20**, 101 (1957)

Chapter 5
5. L. ten Gate. *A note on a simple and rapid method of bacteriological sampling by means of agar sausages* J. Appl. Bact. **28**, 221 (1965)

Chapter 9
6. Report of the Min. of Health (1951) *The purification of the water of swimming baths.* H.M.S.O.
7. Report No. 71. Min of Health (1956). *On the bacteriological examination of water supplies.* H.M.S.O.

Chapter 7 and Chapter 10
8. Technical Bulletin No. 17 Min. Ag. Fish and Food. *Bacteriological techniques for dairy purposes.* H.M.S.O.

Chapter 8
9. Elek, S.D., & Milson, G.F.R. *Combined agar diffusion and replica plating techniques in the study of antimicrobial substance.* J. Clin. Path. **7**, 37 (1954).